DISCOVERING CALCULUS WITH THE GRAPHING CALCULATOR

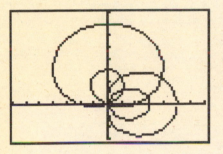

MARY MARGARET SHOAF-GRUBBS, Ph.D.
COLLEGE OF NEW ROCHELLE

JOHN WILEY & SONS, INC.
New York   Chichester   Brisbane   Toronto   Singapore

ISBN 0-471-00974-1

Printed in the United States of America

10 9 8 7 6 5 4 3 2 1

Printed and bound by Malloy Lithographing, Inc.

To
Henry, Jacob, Jim, Tom, and Paul
and, of course, Penny and Casey

Preface

Discovering Calculus With the Graphing Calculator has been written as an enrichment supplement to a course in one-variable calculus. Although this lab manual will nicely supplement most traditional calculus books, the labs roughly follow the flow of the fifth edition of *Calculus* by Howard Anton, and the seventh edition of *Calculus One and Several Variables* by S. L. Salas and Einar Hille and revised by Garret J. Etgen. Graphs are a very important component of calculus and play a significant role in helping students gain a better understanding of calculus. The graphing calculator provides students with an extremely powerful tool to aid in this understanding and helps to provide insight into traditional calculus topics through graphical representations. Furthermore, it is a tool that is controlled by the student himself/herself. It provides students with a means of concrete imagery and gives each student a new control over his/her learning environment and the pace of that learning process. This manual has been written to encourage conceptual understanding in the place of rote memorization. Furthermore, because this conceptual understanding building process is controlled by the student it enables him/her to build a solid understanding of the concept which could then lead into abstraction of the concept.

To the Instructor

The material in the laboratory exercises is not intended to replace a calculus textbook, but, rather, it is intended to help students to apply calculus concepts with a better and more complete understanding. All of the exercise problems can be done on most graphing calculators. The screens shown in this manual are from the TI-82 graphing calculator for the simple reason that I use that model in my own class. These exercises are a product from my own experience and research with graphing calculators in the calculus classroom. The following are suggestions for incorporating technology into your own classroom.

- Do not feel overwhelmed by the graphing calculator technology. You may not want to incorporate each lab exercise into your course. In some of the longer labs, you may decide to assign only some of the problems.

- Encourage students to work in groups of two, three, or four both in and out of the classroom. If possible, include time during class to do some of the lab exercises.

- Encourage students to talk about their observations even if they are not totally correct. Not only do students learn from their mistakes and misconceptions, but teachers have an ideal opportunity to hear what their students are actually thinking and to help correct any misconceptions.

- Assign lab exercises along with textbook exercises. Most exercises ask students to confirm their graphical conclusions analytically or support their analytical results graphically. Students should become familiar and comfortable with the numerical, analytical, and graphical representation of a concept. This helps them to develop a deeper conceptual understanding of the mathematics being taught.

- You may want to assign certain lab exercises as the take-home part of an exam to be handed in at the time of the in-class exam. Not only does this give you the opportunity to examine each student's lab work, but it also will help the student in preparing for the in-class exam. Then write in-class examinations that provide your students with the opportunity to demonstrate their understanding of the concepts.

- Remember that you will be learning along with your students. Technology provides insights into mathematical concepts not only for the students, but also for teachers. Do not be apprehensive about not knowing the answer to each and every question that may come up in class due to the use of technology. Be willing to learn along with your students!

Mary Margaret Shoaf-Grubbs, Ph.D.
mmshoaf@aol.com

Table of Contents

Pitfalls in Using Technology

Technology gives us some very powerful tools with which to study and explore mathematical concepts. One of the most affordable and useful of these tools is the graphing calculator. Many practical applied problems in calculus would be extremely time consuming and difficult to solve using analytical paper-and-pencil methods. There are problems in calculus which cannot be solved analytically. Technology in the form of computer algebra systems (CAS) and the graphing calculator now gives us the means to solve these problems. There are, however, limits to technology and students must be aware of these limits in order to judge whether the result given by a graphing calculator does, indeed, make sense. Some of these pitfalls in using technology will be discussed briefly in the following pages. When reviewing the result given by the graphing calculator students must always ask themselves the following question. Does this graph/answer make sense? If not, then it is possible that you have encountered one the pitfalls in graphing calculator technology.

Round-off and Cancellation Errors

The number of significant digits in a decimal number is the number of digits between the first and last nonzero digit. Calculators can handle only a specified number of digits. The manual for your graphing calculator would tell you the number of digits your particular calculator handles. Calculators either round off the number to the specified number of digits or truncate (drop off) the digits they cannot handle. When many numbers are used in a calculation, the cumulative effect of many small round off errors can be large. One such example in which you will encounter this type of cumulative effect will be in the Laboratory Exercise in Chapter 4 where you will be using Newton's Method to approximate a solution of an equation $f(x) = 0$. The calculator will round off the answer and you will be using that answer to compute the next answer. Thus, the small errors will lead to larger errors.

Round off errors can also lead to cancellation errors. These errors occur in computations of small differences between large quantities. Cancellation errors often produce an error of magnitude 10. Numbers that are different will be the same after rounding.

Overflow and Underflow Errors

Each calculator can handle only a specific number of digits. If a number is larger than the calculator can handle, the error is called overflow. If the number is too small for the calculator to store, the error is called underflow. When a small number is divided by a large number, the answer many be smaller than the calculator can store. The calculator will then store this number as zero. This type of problem occurs in use of technology, but it is more noticeable in calculators that can handle only a small number of digits.

Choosing Friendly Windows in Order to Obtain a More Accurate Picture/Graph

You will be using your graphing calculator to investigate the behavior of many functions. Your calculator *may* show weird looking graphs with false asymptotes if your viewing rectangle is not a 'friendly window'! You must be careful to set your particular calculator on a friendly screen so that conclusions that you make from examining the graph of a function will not be incorrect. As an illustration for the TI-82, consider the following function in two different viewing rectangles (grids).

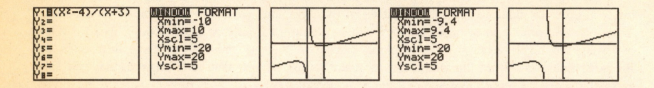

The second of the two function representations is correct is because the magic number for the TI-82 is 94---there are 94 horizontal spaces (95 pixels) on the TI-82. As long as you know the number of horizontal pixels on your particular graphing calculator, you should be able to find friendly windows fairly easily. However, if you are not careful with your choice of viewing rectangles a weird picture like the first graph above could mislead you. It is also possible to avoid false asymptotes by using **DOT MODE** instead of **CONNECTED MODE**. Consult your calculator's manual to find the number of horizontal pixels your particular calculator has.

Another example of how important it is to be in control of your viewing rectangle and exactly what it is that the graph *should* be telling you has to do with the concept of continuity. Consider the screens and graphs from the TI-82 graphing calculator given below.

Examine the function given below in two different viewing rectangles as *x approaches 2*.

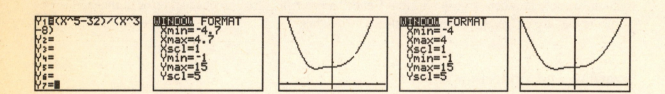

Notice that the first viewing rectangle *does show* that the y-value at $x = 2$ is **undefined**. You can see the gap or empty space in the graph at $x = 2$. However, the second viewing rectangle that differs only in the slightest *does not show or imply this at all!*

Using Built-in Features/Utilities of Graphing Calculators Carefully

Graphing calculators are coming with more and more built-in features such as **PROGRAM**ming capabilities, **DRAW**ing features that include the ability to draw horizontal, vertical, and tangent lines, and, of course, the ability to graph and/or evaluate derivatives and integrals. It is necessary to use these features wisely and *never* accept the result as the as the completely correct answer. As an example, consider the following case.

With some graphing calculators, care must be taken when using the **DRAW INVERSE** utility. As an example, consider the even function $y = x^2$. It is an even function and fails the Horizontal Line Test. However, the TI-82 will draw in inverse of this even function *even though it does not exist!* The calculator draws the image of the parabola reflected about the line $y = x$! You know this cannot be possible, because $y = x^2$ is an even function! Yet, the calculator does, indeed, draw it even though an inverse does not exist! The calculator's results are shown in the first row of calculator screens shown on the next page. Another example is the function $y = \tan x$. The second row of calculator screens shown on the next page shows you how the calculator **DRAW**s the **INVERSE** of this function.

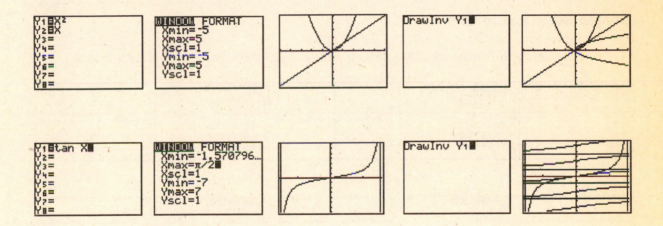

Graphing Conics in Squared Viewing Windows

When graphing circles and ellipses (and also perpendicular lines) you will need to have a *squared viewing window* in order to represent the conics appropriately. A squared viewing window is one in which the distance between the tick marks on the x-axis is the same as the distance between those on the y-axis. To illustrate consider the following graphs of an ellipse found on the next page.

Circle the graph that most <u>appropriately</u> represents this ellipse as we are used to seeing it?

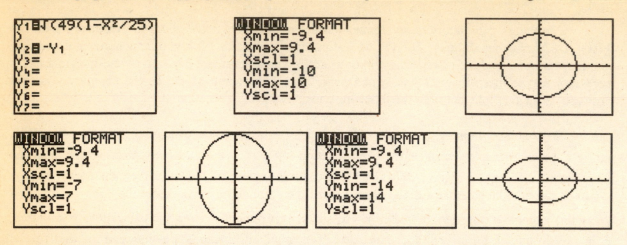

While it is true that all three graphs are correct for the given ellipse, only one represents it the way we are accustomed to seeing it represented.

Hidden Behaviors in Graphs

Consider the following function and its derivative as given in the viewing screens below.

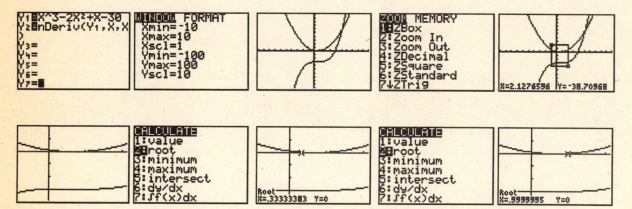

This is a beautiful example of a hidden local maximum and a hidden local minimum!

There are many more pitfalls that you will encounter while using technology to explore mathematical concepts. You must always be aware of this and constantly ask yourself the question: Does what I am seeing make sense?

Chapter 1

Functions and Graphs----A Review

Laboratory Exercise 1.1
Graphs of Inequalities

Name _____   Due Date _____

1.   Solve the following inequality algebraically: $2x - 1 < 11x + 9$
Graph and label each side of the inequality on the provided set of axes (grid) and indicate those values
of x where the line $y = 2x - 1$ is below the line $y = 11x + 9$.

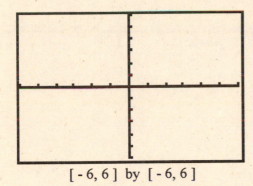

[- 6, 6] by [- 6, 6]

Find the point of intersection algebraically and label it on your graph. Show all work in the space
provided above. If your graphing calculator has an **INTERSECT** utility, confirm your point of
intersection solution obtained through algebra using this utility. If you do not have this feature on your
graphing calculator, use **ZOOM-IN** to confirm your answer.

2.   Solve the following inequality algebraically: $- 1 > 3 - 8x > - 10$
Graph and label each part of the inequality on the provided set of axes (grid) and indicate those values
of x that make the inequality true.

[- 2, 2] by [- 12, 1]

Find the points of intersection algebraically and label them on your graph. If your graphing calculator has an **INTERSECT** utility, confirm your points of intersection solutions obtained through algebra using this utility. If you do not have this feature on your graphing calculator, use **ZOOM-IN** to confirm your answers. Show all work in the space provided above.

3.   Solve the following inequality using your graphing calculator: $x^3 - 3x + 2 < 0$.

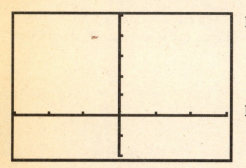

Find values of x that make the inequality **true**.

Find values of x that make the inequality **false**.

[- 3, 3] by [- 2, 5]

4.   Find all values of x for which the expression $\sqrt{x^2 + x - 6}$ can be evaluated. Confirm your results by graphing the expression on the provided grid.

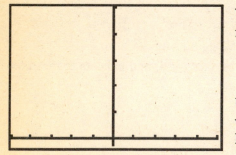

Why does the graph appear to 'stop/quit' for certain values of x?

Use the **TRAC**ing feature of your calculator. **TRACE** to values of $- 3 < x < 2$. What are the values of y for these x's? Explain why this is true.

[- 0.25, 5] by [- 5, 5]

2

5.   In many cases it is difficult to solve an equation or inequality analytically.   Using a combination of analytical and graphical knowledge can frequently help in finding the solution(s) to a difficult problem.   Graph the function on the grid provided below:

$$f(x) = \tan(x/2) + 2x \qquad \text{over the interval } [-2\pi, 2\pi].$$

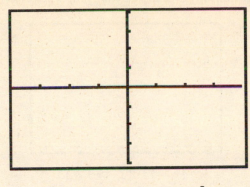

$[-2\pi, 2\pi]$ by $[-20, 20]$

Now solve the inequality $-2x \leq \tan\left(x/2\right)$ using what you know about algebra and trigonometry along with the above graph.   Keep in mind the domain of the tangent function.

Is the result you obtained above confirmed when you graph the inequality?   (Graph the inequality as two individual equations and examine the resulting graph to determine where the line -2x is below the tangent function.

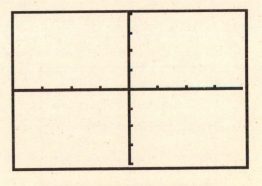

$[-2\pi, 2\pi]$ by $[-20, 20]$

3

6.   (a)   Use your knowledge of algebra and inequalities to find all values of x that satisfy the conditions:

$$(x + 2)(x + 1) > 0 \qquad \text{and} \qquad \frac{x + 2}{x + 1} > 0 \ .$$

Support your results graphically and sketch your graph below on the following grid.

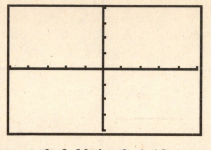

[- 5, 5] by [- 4, 4]

(b)   Now change the addition signs to subtraction signs.   Find the solutions to the inequalities and support your results graphically.

$$(x - 2)(x - 1) > 0 \qquad \text{and} \qquad \frac{x - 2}{x - 1} > 0$$

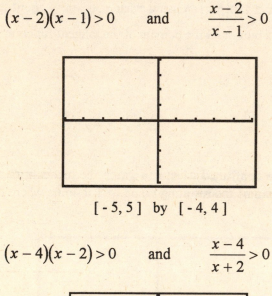

[- 5, 5] by [- 4, 4]

$$(x - 4)(x - 2) > 0 \qquad \text{and} \qquad \frac{x - 4}{x + 2} > 0$$

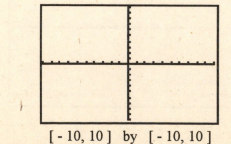

[- 10, 10] by [- 10, 10]

Do you see a pattern occurring in the three inequalities? Form a conjecture about your observations and test it with some inequalities of your own.

4

Laboratory Exercise 1.2
Graphs of Absolute Values

Name _____   Due Date _____

1.   Use your graphing calculator to graph each of the following functions. Sketch and label each graph on the provided grids. Explain how the line, *f(x)*, is changed when absolute value becomes part of the function as in *g(x)* and *k(x)*. Write your conclusions below.

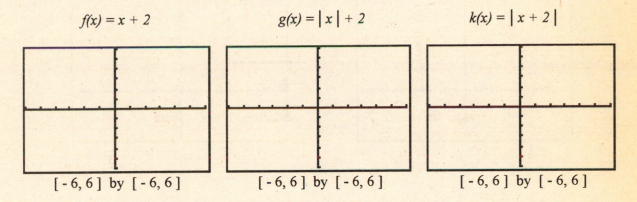

$f(x) = x + 2$ $g(x) = |x| + 2$ $k(x) = |x + 2|$

[- 6, 6] by [- 6, 6]   [- 6, 6] by [- 6, 6]   [- 6, 6] by [- 6, 6]

2.   Graph the function $f(x) = 9 + 6x - x^2$ on the first grid provided below. Using this graph predict what the graph $g(x) = |9 + 6x - x^2|$ will look like. Explain your reasoning. Then graph *g(x)* on the second grid to confirm your prediction. If your prediction was not the same as the graph of *g(x)*, then re-think your prediction and modify your reasoning.

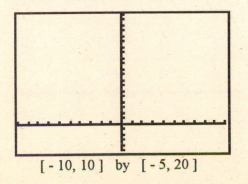
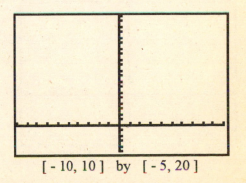

[- 10, 10] by [- 5, 20]   [- 10, 10] by [- 5, 20]

3.   Graph the function $f(x) = x^3 - x^2 - 4x + 6$ on the first grid provided below. Using this graph predict what the graph $g(x) = | x^3 - x^2 - 4x + 6 |$ will look like. Explain your reasoning. Then graph g(x) on the second grid to confirm your prediction. If your prediction was not the same as the graph of g(x), then re-think your prediction and modify your reasoning.

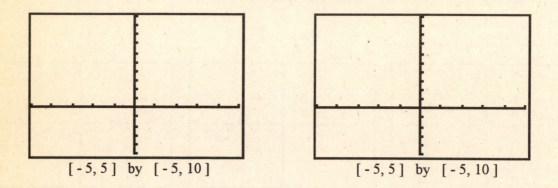

[- 5, 5] by [- 5, 10]   [- 5, 5] by [- 5, 10]

4.   Graph the function $f(x) = x^3 - x^2 - 4x + 4.8$ on the first grid provided below. Using this graph predict what the graph $g(x) = | x^3 - x^2 - 4x + 4.8 |$ will look like. What term of the cubic makes this problem more difficult to predict then the cubic in #3 above? What feature of your graphing calculator will help you to predict the graph of g(x) more accurately? Explain your reasoning. Then graph g(x) on the second grid to confirm your prediction. If your prediction was not the same as the graph of g(x), then re-think your prediction and modify your reasoning.

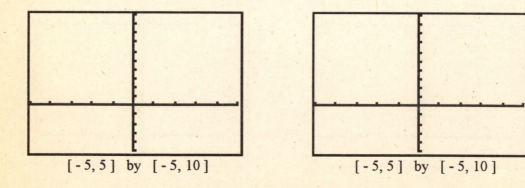

[- 5, 5] by [- 5, 10]   [- 5, 5] by [- 5, 10]

5.   Find all values of x that make $|3x - 7| = 3$ true. Show all of your work in the space below. Confirm your algebraic solution by graphing the two sides of the equation and sketching the results on the provided grid. Indicate on the graph those values of x that are solutions to the equation. Explain your answer.

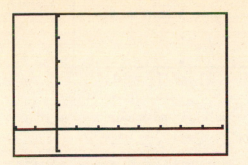

[-2, 8] by [-1, 5]

6.   Find the solution to $|x + 2| = |x - 1|$. Show your work below and confirm your algebraic solution by graphing each side of the equation on the grid below. Indicate on the grid those values of x that make the equation true.

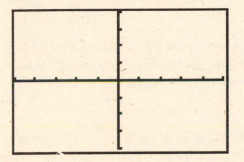

[-5, 5] by [-4, 4]

What do you notice about these absolute value graphs? Look back to Problem 1 and modify this problem so that there are **no solutions**? Is there only one way in which to accomplish a result of no solution? Can you modify the problem so that there are 2 solutions? Explain.

7.   Graph the equation $f(x) = |2 - x|$ and sketch it on the first grid provided on the next page. At what values of x does f(x) equal $2 - x$? At what values of x does f(x) equal $x - 2$? Confirm your predictions by graphing $2 - x$ and $x - 2$ on the second grid.

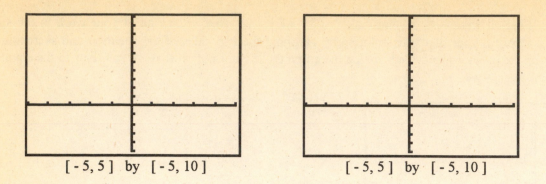

$[-5, 5]$ by $[-5, 10]$ $[-5, 5]$ by $[-5, 10]$

8.   Find the solution to $|x - 2| = 2x + 1$. Show your work below and confirm your algebraic solution by graphing each side of the equation on the grid below.

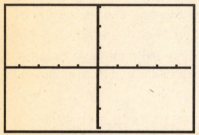

$[-4.7, 4.7]$ by $[-3, 3]$

If your algebraic solution was incomplete you may have obtained two results for the value of x. Does your graph confirm **both** of those results as being solutions for this equation? Why or why not? What is the name for this type of root?   Find an equation for which this extraneous root is a root!

9.   Examine the graph of the absolute value function given on the grid below. It is possible to write this function in the form of a piece-wise function in which the absolute value function does not appear. It is also quite possible to obtain the piece-wise function by using information from the given graph and by-passing the algebraic method. See if you can find the new representation of this function in this way. Show all of your work. Check your results by graphing your resulting equation on the same viewing rectangle as the original function.

$$f(x) = |x - 1| + |x - 3| + 2|x - 4| + 3|x - 5|$$

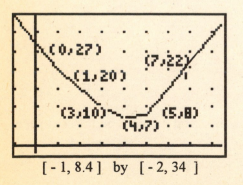

$[-1, 8.4]$ by $[-2, 34]$

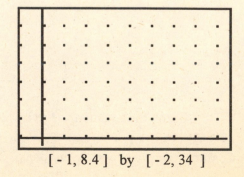

$[-1, 8.4]$ by $[-2, 34]$

Laboratory Exercise 1.3
Slope, y-Intercepts, and the Equations of Lines

Name _____   Due Date _____

1.   Graph each of the following lines on the same grid provided below and label each line appropriately. What do each of the lines have in common? How do they differ?

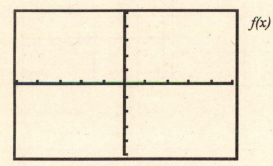

$f(x) = 3x - 2$ $g(x) = 3x + 2$ $k(x) = 3x - 5$

Pick two points on one of the lines on the graph. Find the slope of that line by using the coordinates of the two points. Now repeat the same process for each of the other two lines. Explain your results.

[-5, 5] by [-5, 5]

2.   Graph each of the following lines on the same grid provided below and label each line appropriately. What do each of the lines have in common? How do they differ?

$f(x) = x - 2$ $g(x) = (2/3)x - 2$ $k(x) = 0.25x - 2$

[-5, 5] by [-5, 5]

3.   Graph **each** of the following lines on the two grids provided on the next page and label each line appropriately. Why do they appear different?

$f(x) = 4x - 1$ $g(x) = (-1/4)x - 2$

9

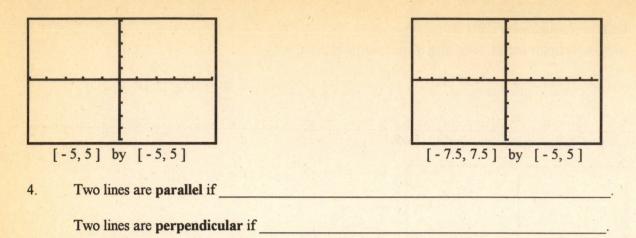

[- 5, 5] by [- 5, 5]   [- 7.5, 7.5] by [- 5, 5]

4.   Two lines are **parallel** if _____.

        Two lines are **perpendicular** if _____.

Many graphing calculators have non-square rectangular screens. If a grid of [-10, 10] by [-10, 10]
is used to graph perpendicular lines, will the lines appear to be perpendicular on the screen? Explain
your answer. Graph two lines that you know are perpendicular. Find a grid dimension that will 'show'
the lines being perpendicular on your screen. How did you arrive at your grid dimensions that 'show'
your lines to be perpendicular?

5.   Write an equation for each of the following lines in Slope-Intercept Form. Give the slope and
y-intercept of each line.

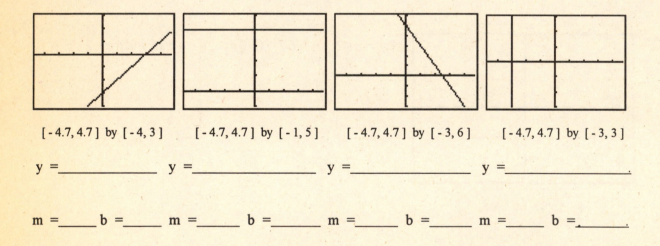

    [- 4.7, 4.7] by [- 4, 3]   [- 4.7, 4.7] by [- 1, 5]   [- 4.7, 4.7] by [- 3, 6]   [- 4.7, 4.7] by [- 3, 3]

y =_____   y =_____   y =_____   y =_____.

m =____ b =____   m =____ b =_____   m =____ b =_____   m =____ b =_____.

10

6.   Find an equation of the line that passes through the point (4, - 3) and has a slope of 2/5. Use your graphing calculator to confirm your result and sketch the graph on the grid provided below. How can the **TRAC**ing feature help you to check your result? (**SUGGESTION**: You might find it helpful to work with the **GRID ON** feature if your calculator has that feature.)

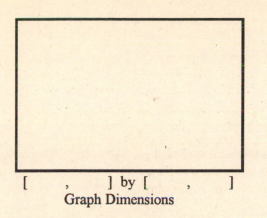

[,   ] by [,   ]
Graph Dimensions

7.   The following three graphs show the line y = 5x + 3. Each of the tick marks on the graphs indicates one unit. Explain why the slopes of the lines appear to be different.

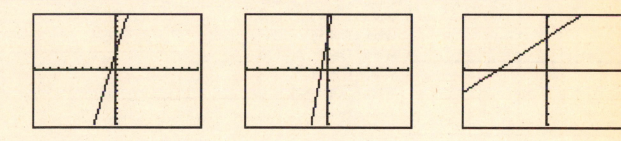

11

8. It is often necessary to convert from Celsius C temperature to Fahrenheit F temperature. One linear equation for this relationship can be given as $F = \dfrac{9}{5}C + 32$. Graph this function on the provided graph and use the graph to find the corresponding temperatures for those given.

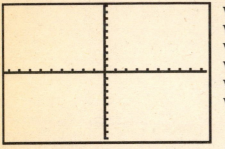

[- 94, 94] by [- 100, 100]

When F = 18°, C = _____ .
When F = 62°, C = _____ .
When F = 156°, C = _____ .
When C = 118.4° ,F = _____ .
When C = −40°, F = _____ .
When C = −7.6°, F = _____ .

Now find the equation for C in terms of F. Find an appropriate viewing window for your new function and then find the corresponding temperatures for those given.

When F = 30°, C = _____ .
When F = −36°, C = _____ .
When F = 56°, C = _____ .
When C = 8°,   F = _____ .
When C = −42°, F = _____ .
When C = −62°, F = _____ .

[,   ] by [,   ]
    Graph Dimensions

12

Laboratory Exercise 1.4
Distance

Name _____   Due Date _____

1.   Find the distance between the points (-4, 3) and (7, - 6). _____

2.   Prove that the following triangle is a right triangle. Show all of your work and explain your results.

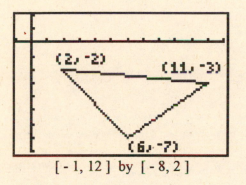

[- 1, 12] by [- 8, 2]

Can you think of another way to prove that this is a right triangle? Explain your reasoning.

3.   Find the point (2, *k*) so that the triangle with vertices (1, 2), (3, 2), and (2, *k*) is equilateral. Is there more than one answer? Sketch in the triangle(s) on the provided grid. Show all work.

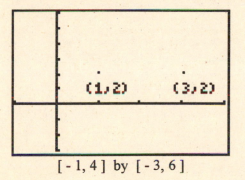

[- 1, 4] by [- 3, 6]

Do(es) the triangle(s) appear to be equilateral? Explain why or why not.

Laboratory Exercise 1.5
Equations of the Form $y = ax^2 + bx + c$

Name _____   Due Date _____

1.   Sketch the following parabolas on the same grid and label appropriately.
      $f(x) = x^2 - 2x - 2$ $g(x) = x^2 - 2x - 6$ $k(x) = x^2 - 2x + 5$

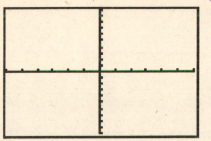

What characteristic(s) do the parabolas have in common?

   [- 6, 6] by [- 10, 10]

2.   Use your graphing calculator to graph the parabola $y = (x - 4)(x + 2)$. Then find the vertex of the parabola. (Recall from algebra that $x = -b/2a$.) You can also use the **ZOOM**, **TRACE**, or **MINIMUM** feature of your graphing calculator! What are the zeros of this parabola?

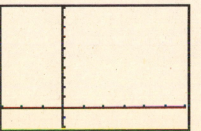

Vertex = _____

Zeros = _____

   [- 6, 6] by [- 10, 10]

3.   Sketch the graph of the parabola $y = x^2 + 4x - 5$ in the grid [- 3, 6] by [- 2, 10] below.

Can you find the roots/zeros of this parabola from its graph as it appears on your calculator screen?   Why not?

Use the Quadratic Formula to find the roots/zeros of this parabola and confirm your results by **ZOOM**ing out on your calculator.

4.   Find a parabola with zeros of 7 and - 5. Write an equation of this parabola in at least two forms. Sketch your results on the grid below. Confirm your zeros and vertex with the **ROOT** and **MINIMUM/MAXIMUM** or the **ZOOM** features of your graphing calculator.

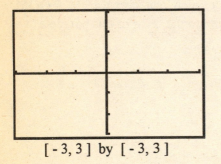

Equations of the parabola: y = _____

y = _____

y = _____

Zeros of the parabola: _____

[- 8, 8] by [- 40, 10]

5.   It is often difficult (or impossible) to find the zeros and/or vertex of a polynomial by analytical means. Use the **ROOT** and **MINIMUM** or the **ZOOM-IN** features of your graphing calculator to find the zeros and vertex of the polynomial $y = 2x^2 - 1.5x - 2.6$. Your results should show that even though the equation does not appear to be that difficult or 'messy with a lot of decimal places' the zeros and vertex are not integer numbers/values.

[- 3, 3] by [- 3, 3]

6.   Give the equation of the transformed graph of $y = x^2$ by performing a horizontal shift of - 4, a vertical shift of 2.5, and a vertical shrink of 0.5.

Equation of the parabola: y = _____

[,   ] by [,   ]

Chapter 2
Limits and Continuity

Laboratory Exercise 2.1
Zeros, Domain, and Range

Name _____   Due Date _____

Use your graphing calculator to investigate the behavior of the following functions. Remember that your calculator *may* show 'weird' graphs with 'false asymptotes' if your viewing rectangle is not a 'friendly window'! (You may need to refer back to the first unit in this book, *Pitfalls of Technology.*) Be careful to set your particular calculator on a 'friendly screen' so that your conclusions will not be incorrect. As an illustration for the TI-82, consider the following function in two different viewing rectangles (grids).

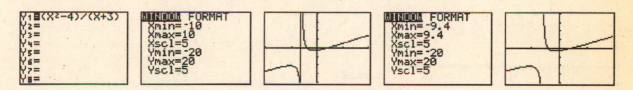

The second of the two function representations is correct is because the 'magic number' for the TI-82 is 94---there are 94 horizontal spaces (95 pixels) on the TI-82. As long as you know the number of horizontal pixels on your particular graphing calculator, you should be able to find 'friendly windows' fairly easily. However, if you are not careful with your choice of viewing rectangles a 'weird picture' like the first graph above could mislead you. It is also possible to avoid false asymptotes by using **DOT MODE** instead of **CONNECTED MODE**.

1.   Sketch the function $f(x) = \sqrt{3x - 10}$ in the indicated grid/window below.

Domain: _____

Range: _____

Zeros: _____

Confirm your graphical solution with the algebraic solutions.

[-1, 10] by [-1, 5]

2. $f(x) = 4x^2 - 3x + 2$

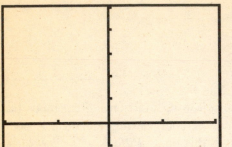

[- 2, 2] by [- 1, 5]

Domain: _____

Range: _____

Zeros: _____

Confirm your graphical solution with the algebraic solutions.

3. $f(x) = x / (x^2 + x + 1)$

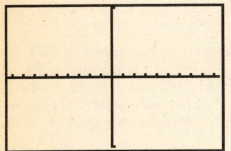

[- 10, 10] by [- 1, 1]

Domain: _____

Range: _____

Zeros: _____

Confirm your graphical solution with the algebraic solutions.

4. $f(x) = \log x$

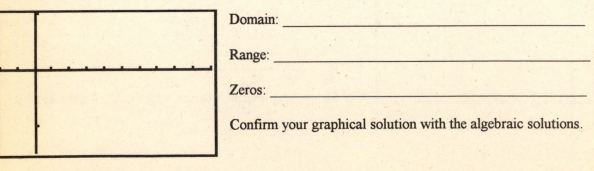

[- 2, 10] by [- 1.5, 1]

Domain: _____

Range: _____

Zeros: _____

Confirm your graphical solution with the algebraic solutions.

18

5. Graph the function $k(x) = \dfrac{3-x}{3-x}$ on the grid provided below. Select your viewing rectangle carefully. Is this the same function as $p(x) = 1$? Why or why not? Explain in detail. Indicate the domain and range of $k(x)$.

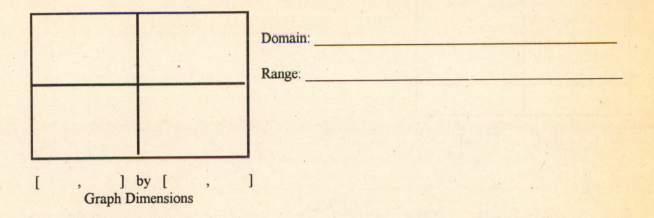

Domain: _____

Range: _____

[,   ] by [,   ]
   Graph Dimensions

6. Graph the function $r(x) = \dfrac{1}{x^3 - 1} + \sqrt{(x+1)(x+3)}$ on the following grid. Indicate the domain and range of the function. Explain your conclusions in detail.

Domain: _____

Range: _____

[,   ] by [,   ]
   Graph Dimensions

7.   Graph the function $f(x) = \dfrac{\sin 4x}{x}$ on the following grid. Indicate the domain and range of the function.

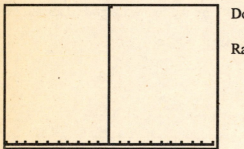

Domain: _____

Range: _____

$\left[-2\pi, 2\pi\right]$ by $\left[-1, 4\right]$

8.   Graph the function $f(x) = \dfrac{2x^2 - x + 3}{3x^2 + 5}$ on the following grid. Indicate the domain and range of the function.

Domain: _____

Range: _____

$[\,-10, 10\,]$ by $[\,0, 1\,]$

20

Laboratory Exercise 2.2
Odd and Even Functions

Name _____   Due Date _____

A function $f(x) = y$ is **even** if $f(-x) = f(x)$ and is **odd** if $f(-x) = -f(x)$ for all x in the domain of f. We also know that an **even** function is symmetric about the y-axis while an **odd** function is symmetric about the origin.

Graph each of the following functions on the grids provided. Decide if the function is even, odd, or neither. Confirm your conclusions algebraically.

1. $f(x) = x^2 - 16$

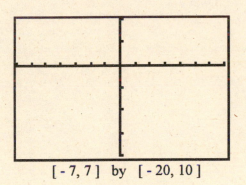

[- 7, 7] by [- 20, 10]

2. $f(x) = \sqrt{x^3}$

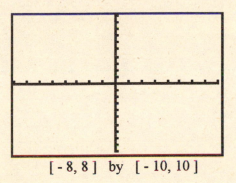

[- 8, 8] by [- 10, 10]

3. $g(x) = x^3 + x$

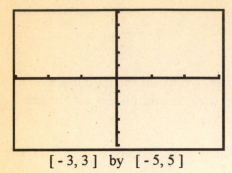

[- 3, 3]   by   [- 5, 5]

4. $k(x) = \dfrac{2}{(x^3 + x)^2}$

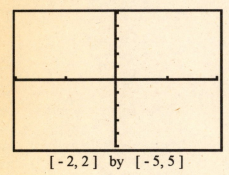

[- 2, 2]   by   [- 5, 5]

5. $k(x) = \dfrac{2}{(x^3 + x)^3}$

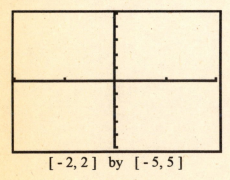

[- 2, 2]   by   [- 5, 5]

Laboratory Exercise 2.3
The Concept of Limit

Name _____   Due Date _____

In this laboratory use your graphing calculator to investigate the limits of the following functions. Sketch the graph of each function on the provided grid. Support your findings analytically. Always keep in mind that a graphing calculator can *only* suggest what the limit is!

1. $\lim\limits_{x \to 3}\ x(\,3 - x\,)$

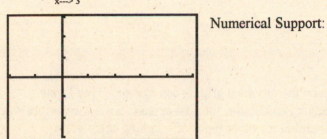

Numerical Support:

     [- 2, 5] by [- 3, 3]

2. $\lim\limits_{x \to 1}\ 3x^2(\,2x - 1\,)$

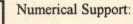

Numerical Support:

     [- 1, 1.5] by [- 1, 4]

3. $\lim\limits_{x \to -1}\ \sqrt{x^3 - 3x - 1}$

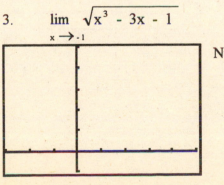

Numerical Support:

     [- 3, 5] by [- 1, 5]

4. $\lim\limits_{x \to \infty} \sqrt{5 + 2x^2} - x$

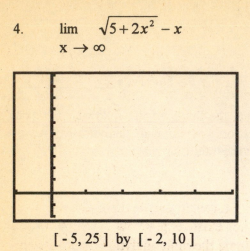

[- 5, 25]   by   [- 2, 10]

5.   Graph the following piecewise function on the provided grid. Then find the given limits. If you are using a TI-81, or TI-82, or TI-85 graphing calculator, the first screen shows the correct way to enter this piece-wise function. Refer to your manual for other types of graphing calculators.

$$f(x) = \begin{cases} \sqrt{1 - x^2} & \text{if } x < 1, \\ 1 & \text{if } 1 \le x < 2, \\ 2 & \text{if } x \ge 2. \end{cases}$$

```
Y1=√(1-X²)(X<1)+
1(1≤X)(X≤2)+2(X≥
2)
Y2=
Y3=
Y4=
Y5=
Y6=
```

[- 1.5, 4]   by   [- 1, 3]

What is the limit near x = 0 ?_____

What is the limit near x = 1 ?_____

What is the limit near x = 2 ?_____

What is the limit near x = 1.5 ?_____

What is the limit near x = - 1? _____

It is possible to determine limits in three ways:

    a.   analytically
    b.   numerically by generating a table of values
    c.   graphically using **TRACE** and **ZOOM** as long as the viewing rectangle is carefully selected.

Of the three methods, only the first gives exact answers while the second and third give approximations to the exact answer.

Use the **ZOOM** feature of your graphing calculator to estimate the limit of the following functions to three decimal positions. If the limit does not exist, explain why. Support your conclusions by constructing a **TABLE** of values. Determine the limit analytically, if possible.

6. $\lim_{x \to 1} \dfrac{2x^3 - 3x + 1}{2x^3 - 3x^2 + 1}$

x	$f(x)$

7. $\lim\limits_{x \to 3} \dfrac{\sin(x-3)}{x+2}$

x	f(x)

8. $\lim\limits_{\theta \to 0} \dfrac{\sin\theta}{\theta^2}$

x	f(x)

9. $\lim\limits_{x \to 0} x^x$

x	f(x)

Does the limit exist? Why not?

Can you think of a simple, yet complete way to say this? See Laboratory Exercise 2.3.

26

Laboratory Exercise 2.4
One-Sided and Two-Sided Limits

Name _____   Due Date _____

Find the given limit, if it exists. If it does not exist, explain why.

1. $\lim\limits_{x \to 2^+} \sqrt{x-2}$

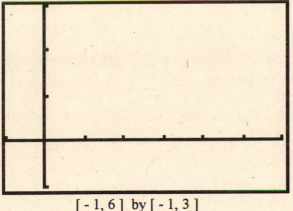

[- 1, 6] by [- 1, 3]

2. $\lim\limits_{x \to 4^-} \dfrac{\sqrt{x-4}}{x+2}$

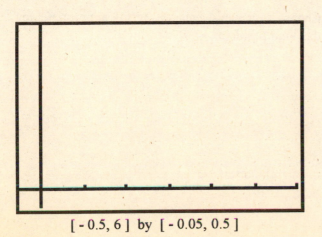

[- 0.5, 6] by [- 0.05, 0.5]

27

3. $\lim\limits_{x \to 4^{+}} \dfrac{\sqrt{x-4}}{x+2}$

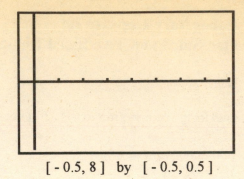

[- 0.5, 8] by [- 0.5, 0.5]

4. $\lim\limits_{x \to +\infty} \sqrt{x^2 + 3} - x$

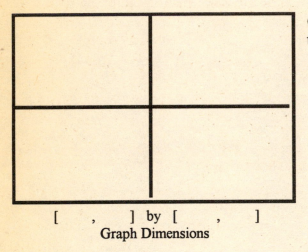

[,   ] by [,   ]
Graph Dimensions

Choose a viewing rectangle (grid) appropriate for this limit problem.

5. $\lim\limits_{x \to +\infty} \sqrt{x^2 + 8} - x$

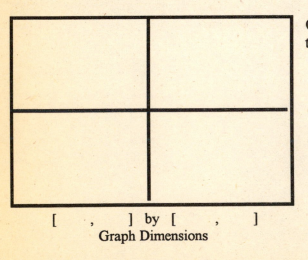

[,   ] by [,   ]
Graph Dimensions

Choose a viewing rectangle (grid) appropriate for this limit problem.

28

6.   Did you notice the similarity of the graphs **and** the limits of Problems 4 and 5? Can you state a conjecture for $\lim\limits_{x \to +\infty} \sqrt{x^2 + a} - x$ based on your observations? You might want to try other values of a before you make a conjecture.

7. $\lim\limits_{x \to +\infty} \sqrt{x^2 + 4x} - x$

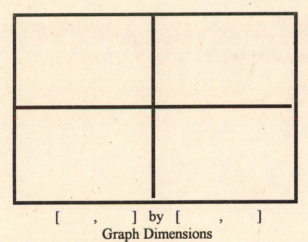

[,   ] by [,   ]
        Graph Dimensions

Choose a viewing rectangle (grid) appropriate for this limit problem.

8. $\lim\limits_{x \to +\infty} \sqrt{x^2 + 7x} - x$

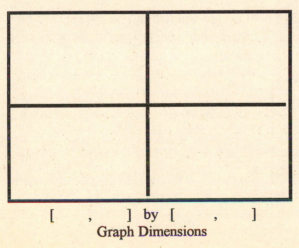

[,   ] by [,   ]
        Graph Dimensions

Choose a viewing rectangle (grid) appropriate for this limit problem.

29

9.   Did you notice the similarity of the graphs **and** the limits of Problems 7 and 8? Can you state a conjecture for $\lim\limits_{x \to +\infty} \sqrt{x^2 + ax} - x$ based on your observations? You might want to try other values of a before you make a conjecture.

10. $\lim\limits_{x \to 0^-} (x^4 + 13x^3 - 16x + 2)$

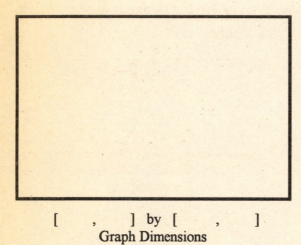

Choose a viewing rectangle (grid) appropriate to show a complete graph for this limit problem.

[,   ] by [,   ]
              Graph Dimensions

5.   Use your graphing calculator to find the $\lim\limits_{x \to 1} \dfrac{x^n - 1}{x - 1}$ both graphically and by generating a table of values. Explore various values of x both integer and non-integer. Be sure to take into consideration both positive and negative values of n. State your conclusions in the form of a conjecture.

n	*x*	*f(x)*

30

Laboratory Exercise 2.5
Defining Limits Formally With Delta and Epsilon
by Examining Input/Output Values

Name _____   Due Date _____

Examine the $\lim_{x \to 3} 3x - 4 = 5$ using your graphing calculator. Suppose we need to "control" x so that y will always be within 1 unit of $y = 5$. We would then need to find x-values such that $4 < y < 6$. This can be visually and numerically explored with the graphing calculator through the following steps. (The view screens shown here are for the TI-82 graphing calculator.)

First examine the function graphically. The **DRAW VERTICAL** feature of the calculator is used to box-in the interval of x-values that gives values of y between 4 and 6.

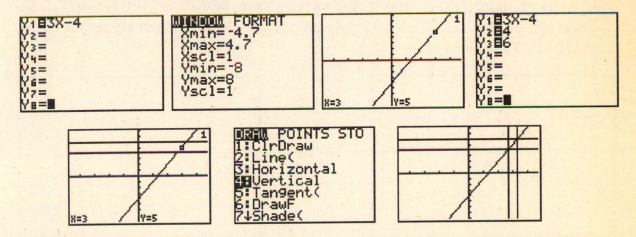

It 'visually appears' that in order for y to be within 1 unit of $y = 5$ that x is somewhere between (approximately) 2.5 and 3.5. Now examine this numerically using the **TABLE** on your calculator.

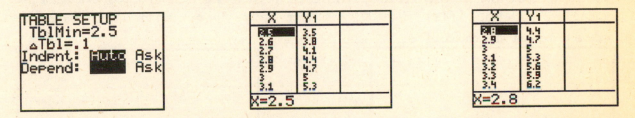

With the **TABLE INCREMENT** set at 0.1, it now appears that the x-values are somewhere between 2.6 and 3.4. Continue to use the **TABLE** to numerically **ZOOM-IN** around y between 4 and 6 by setting the increment to 0001 and 0.001.

31

X	Y1			X	Y1			X	Y1			X	Y1
2.65	3.95			3.28	5.84			2.662	3.986			3.33	5.99
2.66	3.98			3.29	5.87			2.663	3.989			3.331	5.993
2.67	4.01			3.3	5.9			2.664	3.992			3.332	5.996
2.68	4.04			3.31	5.93			2.665	3.995			3.333	5.999
2.69	4.07			3.32	5.96			2.666	3.998			3.334	6.002
2.7	4.1			3.33	5.99			2.667	4.001			3.335	6.005
2.71	4.13			3.34	6.02			2.668	4.004			3.336	6.008
X=2.71				X=3.34				X=2.668				X=3.33	

Numerically x appears to be between 2.666 and 3.333. By using the **2nd CALC INTERSECT** calculator feature, we find that in order for y to be within 1 unit of 5 then $8/3 < x < 10/3$. Putting this in more formal terms we can say the following:

$$| f(x) - 5 | < 1 \quad \text{whenever} \quad 0 < |x - 3| < \frac{1}{3} .$$

Now try the following problems graphically, numerically, and analytically.

1.   Use your graphing calculator to support $\lim\limits_{x \to 1} 3x^4 + x^2 - 7 = -3$.

Then use the **ZOOM-IN, BOX, TRACE, TABLE,** and/or **SOLVER** features of your calculator find a number δ such that $|f(x) + 3| < 0.1$ whenever $0 < |x - 1| < \delta$. Include any graphs that are needed to explain your choice of δ.

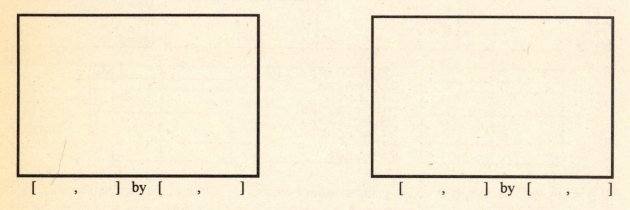

[,   ] by [,   ]   [,   ] by [,   ]

Now use the **ZOOM-IN, BOX, TRACE, TABLE,** and/or **SOLVER** features of your calculator find a number δ such that $|f(x) + 3| < 0.001$ whenever $0 < |x - 1| < \delta$. Include any graphs that are needed to explain your choice of δ.

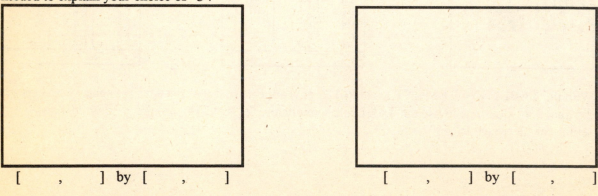

[,   ] by [,   ]   [,   ] by [,   ]

32

2.   Use your graphing calculator to support that $\lim\limits_{x \to -1.4} \sqrt{2 - 5x} = 3$.

Then use the **ZOOM-IN, BOX, TRACE, TABLE,** and/or **SOLVER** features of your calculator find a number δ such that $|f(x) - 3| < 0.01$ whenever $0 < |x + 1.4| < \delta$. Include any graphs that are needed to explain your choice of δ.

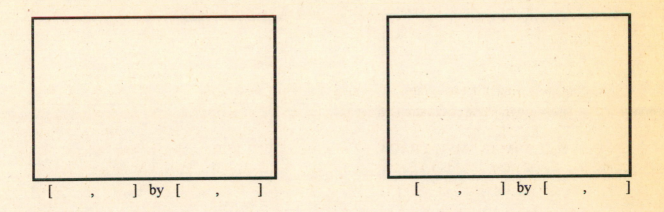

[,   ] by [,   ]   [,   ] by [,   ]

Now use the **ZOOM-IN, BOX, TRACE, TABLE,** and/or **SOLVER** features of your calculator find a number δ such that $|f(x) - 3| < 0.001$ whenever $0 < |x + 1.4| < \delta$. Include any graphs that are needed to explain your choice of δ.

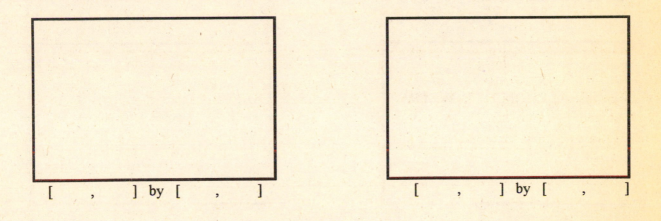

[,   ] by [,   ]   [,   ] by [,   ]

33

3.   Use your graphing calculator to confirm that $\lim\limits_{x \to 3} \dfrac{x^2 - 9}{x - 3} = 6$.

Then use the **ZOOM-IN, BOX, TRACE, TABLE,** and/or **SOLVER** features of your calculator find a number δ such that $|f(x) - 6| < 0.05$ whenever $0 < |x - 3| < \delta$. Include any graphs that are needed to explain your choice of δ.

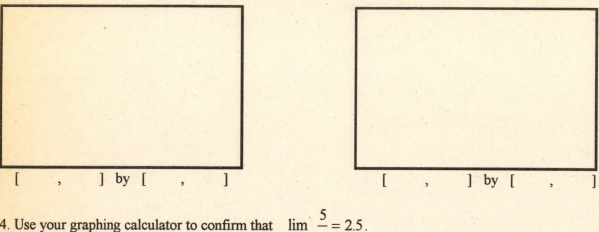

[,   ] by [,   ]   [,   ] by [,   ]

4. Use your graphing calculator to confirm that $\lim\limits_{x \to 2} \dfrac{5}{x} = 2.5$.

Then use the **ZOOM-IN, BOX, TRACE, TABLE,** and/or **SOLVER** features of your calculator find a number δ such that $|f(x) - 2.5| < 0.04$ whenever $0 < |x - 2| < \delta$. Include any graphs that are needed to explain your choice of δ.

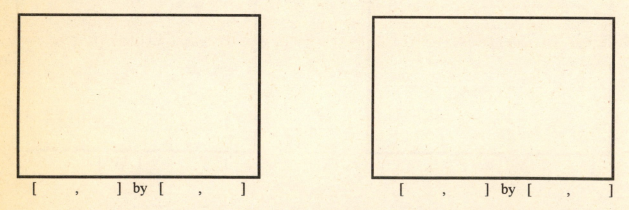

[,   ] by [,   ]   [,   ] by [,   ]

5.   Now re-examine the x- and y-values for the three problems above. Does a small/large change in the x-value result in a small/large change in the y-value? (**NOTE:** The **VERTICAL LINE** feature illustrated in the example is an excellent way to 'see' the graphical changes in x and y while the **TABLE** gives the numerical changes.)

34

Laboratory Exercise 2.6
Continuity

Name _____   Due Date _____

In order for a function $f(x)$ to be continuous at some point, c, certain conditions must be met:

    a. $f(c)$ must exist

    b. $\lim\limits_{x \to c} f(x)$ must exist

    c. $\lim\limits_{x \to c} f(x) = f(c)$

1.   Examine the following graph and answer the questions about continuity at the given values of c.

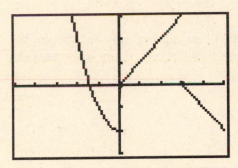

Is f continuous at $x = -1$? Explain.

Is f continuous at $x = 0$? Explain.

Is f continuous at $x = 1$? Explain.

Is f continuous at $x = 3$? Explain.

Is f continuous at $x = 4$? Explain.

$[-5, 5]$ by $[-3, 3]$

2.   Find the points of discontinuity, if any exist, and explain your decision. Graph the function on your graphing calculator and sketch your results below on the provided grid. **REMEMBER:** Select your viewing rectangle carefully! You could also use **DOT MODE** if your calculator has this feature.

$$f(x) = \frac{3x + 1}{x^2 + 7x - 2}$$

[,   ] by [,   ]

3.   Find the points of discontinuity, if any exist, and explain your decision. Graph the function on your graphing calculator and sketch your results below on the provided grid. Confirm your results algebraically.

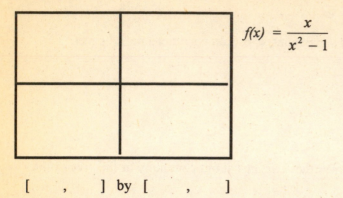

$$f(x) = \frac{x}{x^2 - 1}$$

[,   ] by [,   ]

4.   Determine all points of discontinuity for the given function both algebraically and graphically. Find an appropriate viewing rectangle for the function. Determine which points, if any, are *removable discontinuities*.

$$f(x) = |x + 3| - 4$$

[,   ] by [,   ]

5.   Determine all points of discontinuity for the given function both algebraically and graphically. Find an appropriate viewing rectangle for the function. Determine which points, if any, are *removable discontinuities*.

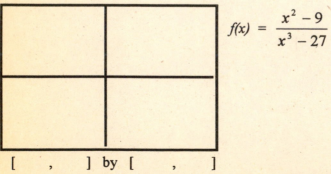

$$f(x) = \frac{x^2 - 9}{x^3 - 27}$$

[,   ] by [,   ]

6.   Determine all points of discontinuity for the given function both algebraically and graphically. Find an appropriate viewing rectangle for the function. Determine which points, if any, are *removable discontinuities*.

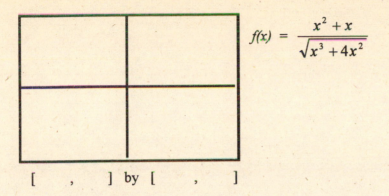

$$f(x) = \frac{x^2 + x}{\sqrt{x^3 + 4x^2}}$$

[,   ] by [,   ]

7.   Graph the function $f(x) = \begin{cases} \dfrac{x^2 - 9}{x - 3} & \text{, if } x \neq 3 \\ 3 & \text{, if } x = 3 \end{cases}$.

Is the function continuous at $x = 3$? Why? If not, then redefine the function at $x = 3$ so that f is continuous at that point.

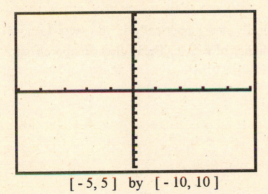

[-5, 5] by [-10, 10]

37

8.   Use your knowledge of functions and composition of functions to examine continuity in the following functions both graphically and numerically. Graph *f(x)* and *g(x)* on the grid provided below.

$$f(x) = x \sin x \qquad \text{and} \qquad g(x) = x^2 + 5$$

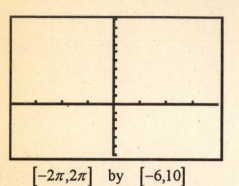

$[-2\pi, 2\pi]$ by $[-6,10]$

Make a conjecture about the continuity of $k(x) = \dfrac{f(x)}{g(x)}$ giving the reasoning behind your conjecture.

Then select an appropriate viewing window and graph *k(x)*.

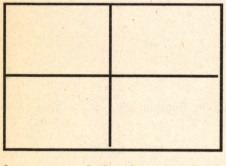

[,   ] by [,   ]
     Graph Dimensions

Make a conjecture about what would happen both graphically and numerically if you were to now graph $m(x) = |k(x)|$. How would this affect the continuity of *k(x)* ? Then select an appropriate viewing window and graph *m(x)*.

[,   ] by [,   ]
     Graph Dimensions

Laboratory Exercise 2.7
Limits and Continuity of Trigonometric Functions

Name _____   Due Date _____

Examine each of the following functions on your graphing calculator to determine the limits and points of discontinuity, if they exist. You may need to view the function in several *different* viewing rectangles in order to feel confident about your conclusions. Sketch the function in the provided grid below and confirm your results analytically. If the limit does not exist, explain why.

1.

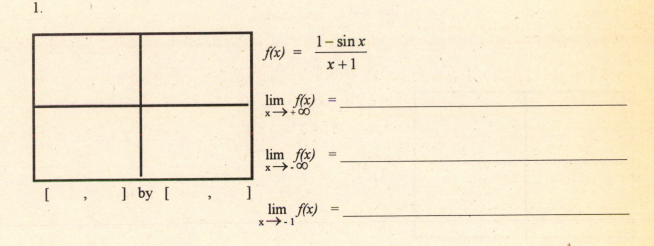

$$f(x) = \frac{1 - \sin x}{x + 1}$$

[,   ] by [,   ]

$\lim_{x \to +\infty} f(x)$ = _____

$\lim_{x \to -\infty} f(x)$ = _____

$\lim_{x \to -1} f(x)$ = _____

2.

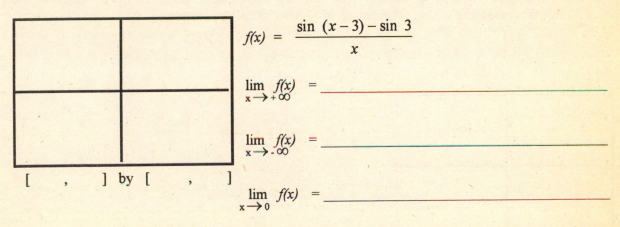

$$f(x) = \frac{\sin (x - 3) - \sin 3}{x}$$

[,   ] by [,   ]

$\lim_{x \to +\infty} f(x)$ = _____

$\lim_{x \to -\infty} f(x)$ = _____

$\lim_{x \to 0} f(x)$ = _____

3.

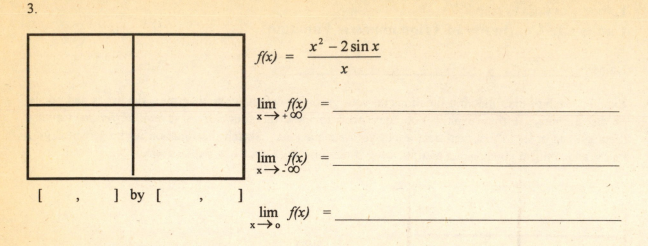

$$f(x) = \frac{x^2 - 2\sin x}{x}$$

$$\lim_{x \to +\infty} f(x) = \underline{\hspace{6cm}}$$

$$\lim_{x \to -\infty} f(x) = \underline{\hspace{6cm}}$$

[,   ] by [,   ]

$$\lim_{x \to 0} f(x) = \underline{\hspace{6cm}}$$

4.

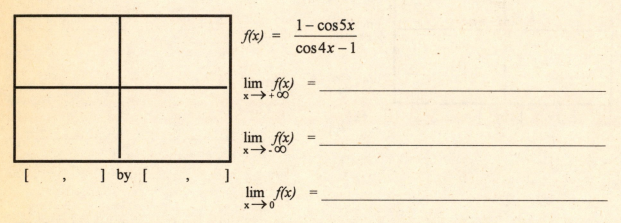

$$f(x) = \frac{1 - \cos 5x}{\cos 4x - 1}$$

$$\lim_{x \to +\infty} f(x) = \underline{\hspace{6cm}}$$

$$\lim_{x \to -\infty} f(x) = \underline{\hspace{6cm}}$$

[,   ] by [,   ]

$$\lim_{x \to 0} f(x) = \underline{\hspace{6cm}}$$

This function is rather 'neat' in appearance. Do you see a pattern in the graph? Does the graph repeat itself? If so, describe clearly what is happening in as much detail as possible.

5.

$$f(x) = \frac{1 + \sin \dfrac{x}{2} x}{x + 1}$$

$$\lim_{x \to -1} f(x) = \underline{\hspace{6cm}}$$

40

6.   Graph the function $f(x) = \sin\dfrac{\pi}{x}$. Use the **ZOOM-IN** and **TABLE** features of your graphing calculator to examine the behavior of f as x approaches 0. State your conclusions and sketch the graph of f on the grid provided below.

[,   ] by [,   ]
Graph Dimensions

Chapter 3

Differentiation

Laboratory Exercise 3.1
Secant Lines and Tangent Lines

Name _____   Due Date _____

If your graphing calculator has **TABLE** and **TABLE SET** features, use them to help you complete the tables in this exercise.

1.   Examine the graph given below of the function $f(x) = x^2$. Investigate the slopes of secant lines between the points (1, 1) and (3, 9) on the curve by completing the following tables of values.

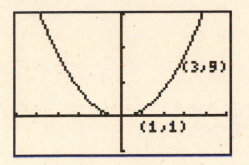

[-5, 5] by [-2, 20]

Table 1		**Table 2**

	x	y
Point 1	3.00	
Point 2	2.75	
Point 3	2.50	
Point 4	2.25	
Point 5	2.00	
Point 6	1.75	
Point 7	1.50	
Point 8	1.25	
Point 9	1.00	

Slope From	
P 1 to P 9	
P 2 to P 9	
P 3 to P 9	
P 4 to P 9	
P 5 to P 9	
P 6 to P 9	
P 7 to P 9	
P 8 to P 9	

What number do the slopes in Table 2 appear to be approaching as the pairs of points get closer and closer together? _____

43

Remember that the *Slope of the Tangent Line* through the point (x_0, y_0) is given by:

$$m_{tan} = \frac{f(x_0 + h) - f(x_0)}{h}$$ as $h \to 0$.   This is referred to as the **Difference Quotient**.

The *Equation of the Tangent Line* can then be written as: $y - y_0 = m_{tan}(x - x_0)$

Find the slope and an equation of the tangent line to the graph of $y = f(x) = x^2$ at $x = 1$.

y = _____

Graph this Tangent Line (TL) equation on the same graph as the original function. Using a 'friendly screen', how can you support your conclusion that TL equation you obtained is, indeed, tangent at $x = 1$?

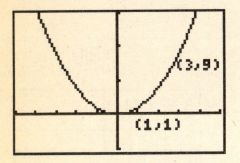

[- 5, 5] by [- 2, 20]

Use the **DRAW → TANGENT LINE (TL)** feature of your graphing calculator (if your calculator has it) to support that your numerical result for the **TL** is the same as the **TL** obtained by your calculator.

44

2.   Examine the graph given below of the function $f(x) = x^2 + 2$.   Investigate the *Slopes of Secant Lines* between points on the curve by completing the following Tables of values.

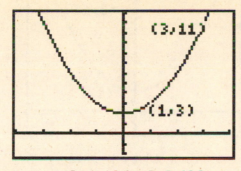

[- 4, 4] by [- 2, 12]

Table 3

	x	y
Point 1	3.00	
Point 2	2.75	
Point 3	2.50	
Point 4	2.25	
Point 5	2.00	
Point 6	1.75	
Point 7	1.50	
Point 8	1.25	
Point 9	1.00	

Table 4

Slope From	
P 1 to P 9	
P 2 to P 9	
P 3 to P 9	
P 4 to P 9	
P 5 to P 9	
P 6 to P 9	
P 7 to P 9	
P 8 to P 9	

What number do the slopes in Table 4 appear to be approaching as the pairs of points get closer and closer together? _____

Are your conclusions here any different from those for Question 1, Table 2 above? Explain.

Find the Slope and an Equation of the Tangent Line to the graph of $y = f(x) = x^2 + 2$ at x = 1.

y = _____

45

Graph this Tangent Line (TL) equation on the same graph as the original function. Using a 'friendly screen', how can you support that your TL equation is, indeed, tangent at $x = 1$?

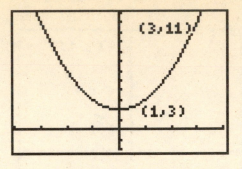

[- 4, 4] by [- 2, 12]

If your graphing calculator has a **DRAW → TANGENT** feature use it to support your numerical result for the TL in order to see if your TL is the same as the TL obtained by your calculator.

4.   Examine the graph given below of the function $f(x) = x^3 - 3$. Investigate the slopes of the secant lines between points on the curve by completing the following tables of values.

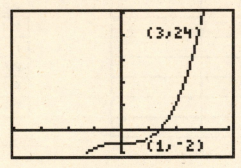

[- 4, 4] by [- 5, 25]

Table 5   **Table 6**

	x	y
Point 1	3.00	
Point 2	2.75	
Point 3	2.50	
Point 4	2.25	
Point 5	2.00	
Point 6	1.75	
Point 7	1.50	
Point 8	1.25	
Point 9	1.00	

Slope	
P 1 to P 9	
P 2 to P 9	
P 3 to P 9	
P 4 to P 9	
P 5 to P 9	
P 6 to P 9	
P 7 to P 9	
P 8 to P 9	

What number do the slopes in Table 6 appear to be approaching as the pairs of points get closer and closer together? _____

Do you think that your results would have been any different if the equation had been $y = x^3$ instead of $y = x^3 - 3$? Explain.

Find the Slope and an Equation of the Tangent Line to the graph of $y = f(x) = x^3 - 3$ at $x = 1$.

y = _____

Graph this Tangent Line (TL) equation on the same graph as the original function. Using a 'friendly screen' how can you confirm that your TL equation is, indeed, tangent at $x = 1$?

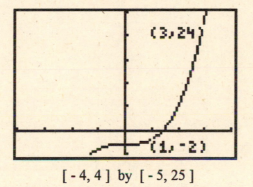

$[-4, 4]$ by $[-5, 25]$

Use the **DRAW → TANGENT** feature of your graphing calculator to support that your numerical result for the TL is the same as the TL obtained by your calculator.

REMEMBER: The limit of the **Difference Quotient** $m_{tan} = \lim \dfrac{f(x_0 + h) - f(x_0)}{h}$ as $h \to 0$ can be interpreted as the *rate of change of f(x)* from x_0 to $x_0 + h$ as h gets smaller and smaller.

47

Laboratory Exercise 3.2
The Difference Quotient and Estimating the Derivative of a Function

Name _____   Due Date _____

The limit of the *Difference Quotient* can now be defined as the *Derivative with respect to x* at the point P(x, y) or:

$$f'(x) = m_{tan} = \lim \frac{f(x+h) - f(x)}{h}$$ as h→0 , provided this limit exists.

1.   (a)   Sketch the graph of $y = f(x) = x^2 + x - 6$ on the grid provided to the right.

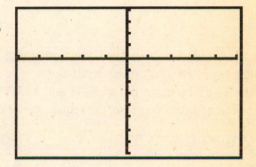

[- 5, 5] by [- 8, 4]

(b)   Use the Difference Quotient (DQ) and small values of h to estimate the derivative of *f(x)* at x = -1.

h	DQ
- 0.01	
- 0.1	
- 0.5	
1	
0.5	
0.1	
0.01	
0.001	
0.0001	

(c)   Choose two or three values of h from the Table above and plot the Difference Quotients against the variable x. Use the grid provided on the next page to sketch in your results. What happens to the graphs of the DQ as h gets smaller and smaller?

49

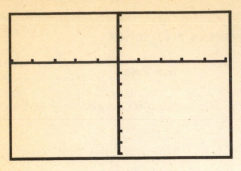

[-5, 5] by [-8, 4]

Estimate the derivative of *f(x)* from your graph(s). Explain how you came to your conclusion.

(d)   Now use your graphing calculator's numerical **Derivative** feature to graph the derivative of *f(x)* on the grid below. The following screens show you how to graph the derivative on the TI-82 by using the **MATH** and **Y-VARS** menus. Have the calculator's active cursor at the **Y2 =** position before you start the sequence of steps shown.

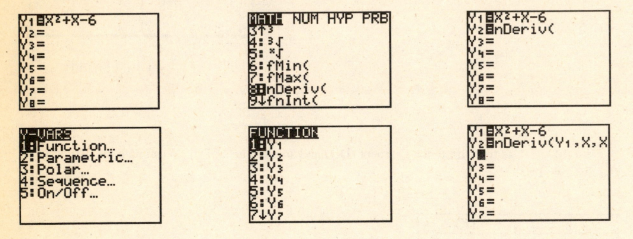

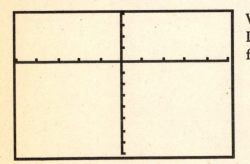

[-5, 5] by [-8, 4]

What do you notice about this graph and the graphs of the Difference Quotient for the various values of h that you found in part (c) above?

Estimate the derivative of *f(x)* at x = -1 using your derivative graph._____

    (e)   On the **HOME SCREEN** of your graphing calculator calculate the numerical derivative of *f(x)* at   x = -1. How does your numerical answer compare with your graphical answer obtained in part (d)?

$f'(-1) =$ _____

2.   (a)   Sketch the graph of $y = f(x) = x^3 - x^2 - 6x$ on the grid provided below.

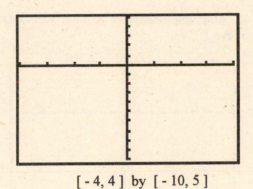

[- 4, 4] by [- 10, 5]

    (b)   Use the Difference Quotient (DQ) and small values of h to estimate the derivative of *f(x)* at x = -2.

h	DQ
1	
0.5	
0.1	
0.01	
0.001	
0.0001	

    (c)   Graph the Difference Quotient for one or more of the values of h in the Table above. Use the grid provided on the next page and sketch in your result(s).

51

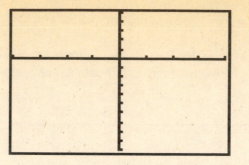

$$[-4, 4] \text{ by } [-10, 5]$$

What happens to the graphs of the Difference Quotient as h gets smaller and smaller?

Estimate the derivative of *f(x)* at x = - 2 from your graph(s). Explain how you came to your conclusion.

(d)   Now use your graphing calculator's numerical **Derivative** feature to graph the derivative of *f(x)* on the grid below.

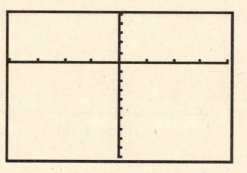

$$[-4, 4] \text{ by } [-10, 5]$$

What do you notice about this graph and the graphs of the Difference Quotient for the various values of h that you found in part (c) above?

Estimate the derivative of *f(x)* at x = -2 using your derivative graph. *f '(- 2)* = _____

(e)   On the **HOME SCREEN** of your graphing calculator calculate the numerical derivative of *f(x)* at x = -2. How does your numerical answer compare with your graphical answer obtained in part (d)?   *f ' (- 2)* = _____

3.   Find the derivative of *f(x)* = $x^3 - x^2 - 6x$ at x = 1.2 from your graph of the function and its derivative, from a **TABLE**, and by using the **HOME SCREEN**.

Laboratory Exercise 3.3
Derivatives That Fail to Exist

Name _____   Due Date _____

In Laboratory Exercise 3.2 the definition of a derivative with respect to x is stated in terms of a limit. The domain of the derivative f' consists of all x's for which the limit exists. What does this actually mean? Graphically this means that the derivative fails to exist where the graph of the function is not smooth. The derivative also fails to exist at all points where the function itself is discontinuous. In this exercise you will examine some graphs and decide where the derivative fails to exist.

1.   Examine and plot the graph of $f(x) = |x - 6| + 3$ on the grid provided below.

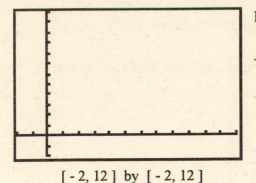

List all points where the derivative of $f(x)$ fails to exist.

_____

[- 2, 12] by [- 2, 12]

2.   Examine and plot the graph of $f(x) = |x^2 - 2x - 3|$ on the grid provided below.

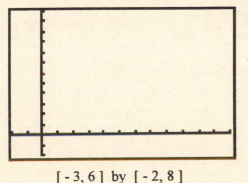

List all points where the derivative of $f(x)$ fails to exist.

_____

[- 3, 6] by [- 2, 8]

53

3. Examine and plot the graph of $f(x) = \dfrac{x+4}{x}$ on the grid provided below.

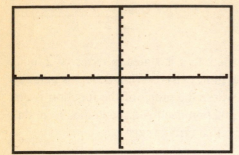

[-4, 4] by [-8, 8]

List all points where the derivative of $f(x)$ fails to exist.

_____

4. Examine and plot the graph of $f(x) = \left| x^3 \right|$ on the grid provided below.

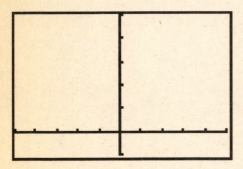

[-5, 5] by [-1, 5]

List all points where the derivative of $f(x)$ fails to exist.

_____

What is $f'(0)$? _____

5. Examine and plot the graph of $f(x) = \begin{cases} x & when\ x < 1 \\ -x^2 & when\ x \geq 1 \end{cases}$ on the grid provided below.

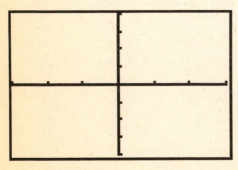

[-2, 12] by [-2, 12]

List all points where the derivative of $f(x)$ fails to exist.

_____

54

6.   Examine and plot the graph of $f(x) = \begin{cases} 2x + 1 & \text{, if } x \leq 2 \\ \dfrac{x^2}{2} + 4 & \text{, if } x > 2 \end{cases}$ on the grid provided below.

What would be an appropriate viewing rectangle?

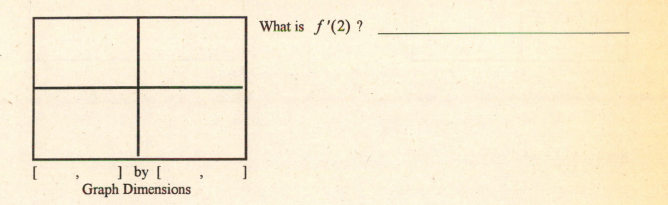

What is $f'(2)$? _____

[,   ] by [,   ]
        Graph Dimensions

Now that you are more familiar with the graphs of different functions, can you create in your own "mind's eye" the graph of each function without actually graphing the function on your calculator? If so, can you determine those values of x where the derivatives of the functions would fail to exist? Explain how you came to those conclusions.

Then try sketching the curves given below. On the first provided grid sketch the function as you see it in your 'mind's eye'. On the second grid sketch in the resulting graph that your calculator obtains. Are there differences? If so, where and why did your graph not look like that of the calculator.

7. $f(x) = x^3 - 2x^2 - 3x$

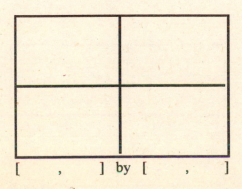

[,   ] by [,   ]

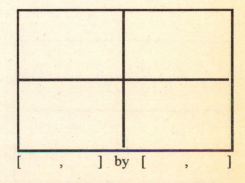

[,   ] by [,   ]

8. $k(x) = 2|x + 4| - 5$

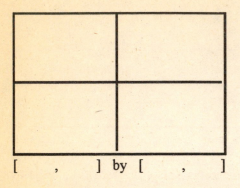

[,   ] by [,   ]

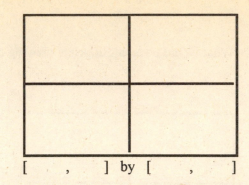

[,   ] by [,   ]

9. $m(x) = x + \dfrac{1}{x}$

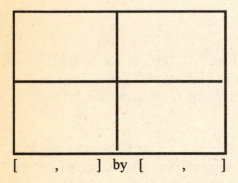

[,   ] by [,   ]

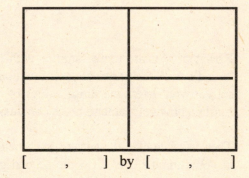

[,   ] by [,   ]

10. $f(x) = \sin 2x$

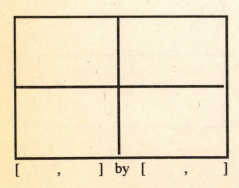

[,   ] by [,   ]

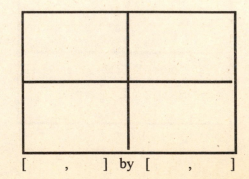

[,   ] by [,   ]

Laboratory Exercise 3.4
Differentiation Rules

Name _____ Due Date _____

Find the derivative for each of the functions given below using the rules of differentiation that you have learned. Support your answers by graphing each function, the derivative that you find, and the numerical derivative that your graphing calculator produces on the same grid. Check your graphing calculator manual to learn the correct way to have the calculator obtain the numerical derivative.

If your derivative result is correct, how should the graphs of the two derivatives (the one you obtained using the rules of differentiation and the numerical derivative the calculator obtained) compare? In some cases it is possible by using differentiation rules to find the derivative in more than one way. Check your derivative result by obtaining the derivative a second way for any of the functions below.

1. $f(x) = (x^2 + 4x - 2)(x^3 - 4)$

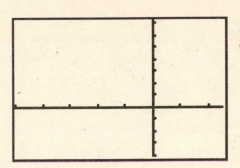

[- 5, 2.5] by [- 100, 180]

Describe the graph of $f(x)$ at each of the x-values where the derivative has a zero.

At what x-values does $f(x)$ have a horizontal tangent?

2. $f(x) = \sqrt{x} \, \sin x$

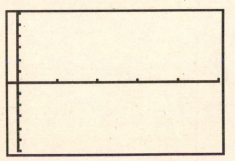

[- 1, 25] by [- 6, 6]

Describe the graph of $f(x)$ at each of the x-values where the derivative has a zero.

At what x-values does $f(x)$ have a horizontal tangent?

3. $f(x) = \dfrac{\sqrt{x}\ \sin x}{x - 3}$

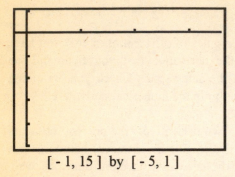

[- 1, 15] by [- 5, 1]

Describe the graph of *f(x)* at each of the x-values where the derivative has a zero.

At what x-values does *f(x)* have a horizontal tangent?

What is $f'(0)$? _____

Why?

4. $f(x) = (x - 3)^2 (x + 2)^3$

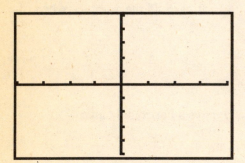

[- 4, 4] by [- 125, 125]

Describe the graph of *f(x)* at each of the x-values where the derivative has a zero.

5.   Examine the following graph of a function and its derivative. Label the graphs appropriately as $f(x)$ or $f'(x)$.

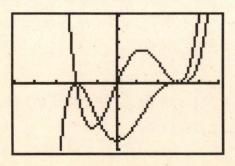

58

Laboratory Exercise 3.5
Velocity and Speed------An Exploration in Parametric Mode
Vertical and Horizontal Motion on the Graphing Calculator

Name _____   Due Date _____

A heavy cannon ball is blasted straight upward with a velocity of 125 ft/sec. The equation for its height in feet after t seconds is: $s(t) = 125t - 16t^2$. Examine the moving cannon ball in **Parametric Mode** in each of the following two ways. After changing the mode of your graphing calculator to **Parametric Mode** enter the set of equations as given below in the calculator screens. The second **Window** screen is a continuation of the first **Window** screen. (NOTE: The number 4 was chosen arbitrarily for X_{1T} . Any number on the horizontal axis could have been selected.)

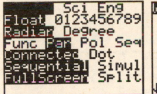

Sketch your results in the grid to the right.

What is the first pair of equations showing you graphically?

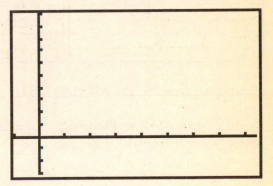

What is the second pair of equations showing you graphically?

What is the maximum height that the cannon ball reaches?_____

How many seconds after it is thrust upward does the cannon ball reach its maximum height?_____

Approximately how many seconds does it take the cannon ball to reach a height of about 186 feet?

_____

Find the equations for the cannon ball's velocity as a function of time and write them in below. Use the third pair of equations on your graphing calculator to graph the ball's velocity.

$X_{3T} = $ _____ $Y_{3T} = $ _____

Find the equations for the cannon ball's speed as a function of time write them in below. Use the fourth pair of equations on your graphing calculator to graph the cannon ball's speed.

$X_{4T} = $ _____ $Y_{4T} = $ _____

QUESTION: For this particular problem, could you have used a TMAX less than 8 seconds? Explain why or why not.

2.   A particle is moving along the path given by the equation: $s(t) = 3T^3 - 10T^2 + 8T$

Enter the first set of equations as given below in the calculator screens.

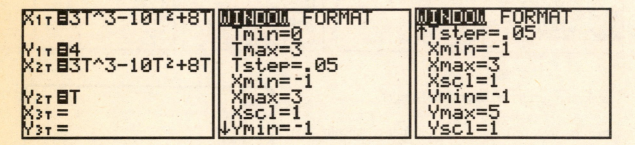

Sketch your results in the grid to the right.

Explain what the first two pairs of parametric equations are telling you graphically.

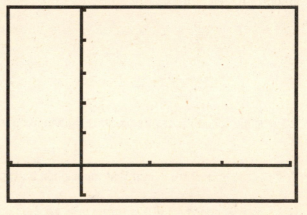

How many times does the particle change direction? _____

At approximately what time(s) does the particle change direction?

60

3.   A particle is moving along the path given by the equation: $s(t) = 3T^3 - 8T^2 + 5T$

Find two parametric representations for the path of the particle. Sketch your results in the provided grid.

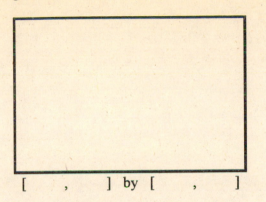

[,   ] by [,   ]

How many times does the particle change direction? _____

At approximately what time(s) does the particle change direction? _____

Chapter 4

Applications of the Derivative

Laboratory Exercise 4.1
Discovering the Relationship Between a Function and its Derivative

Name _____   Due Date _____

1.   Sketch the function $f(x) = x^3 + 3x^2 - 9x + 1$ and its derivative on the grid below.

[- 6, 6] by [- 15, 30]

Describe the behaviors of *f(x)* and its derivative in each of the following intervals.

When $-\infty < x < -3$, *f(x)* _____

_____ and

$f'(x)$ _____.

When $-3 < x < 1$, *f(x)* _____

_____ and

$f'(x)$ _____.

When $1 < x < \infty$, *f(x)* _____

_____ and

$f'(x)$ _____.

2.   Sketch the function $f(x) = (2x^2 - 3x - 2)^2$ and its derivative on the grid below.

[- 3, 3] by [- 15, 15]

Describe the behaviors of *f(x)* and its derivative in each of the following intervals.

When $-\infty < x < -0.5$, *f(x)* _____

_____ and

$f'(x)$ _____.

When $-0.5 < x < 0.75$, *f(x)* _____

_____ and

$f'(x)$ _____.

When $0.75 < x < 2$, *f(x)* _____

_____ and

$f'(x)$ _____.

63

When $2 < x < \infty$, $f(x)$ _____ and
$f'(x)$ _____ .

Make a conjecture about the relationship between these functions and each of their derivatives.

Test your conjecture further with the following functions.   Modify your conjecture if it is necessary.

3.   Sketch the function $f(x) = x^2 + \dfrac{1}{x^2}$ and its derivative on the grid below. Describe the

behavior of *f(x)* and the behavior of its derivative in as much detail as possible. NOTE: It may be necessary to change scale/viewing rectangle in order to see more clearly exactly what is happening in the function and/or its derivative.

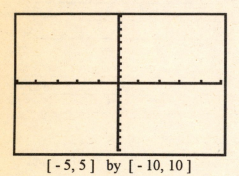

$[-5, 5]$ by $[-10, 10]$

4.   Sketch the function $f(x) = \sqrt[3]{x-1} + \sqrt{x+1}$ for $x \geq -1$ and its derivative on the grid below. Describe the behavior of *f(x)* and the behavior of its derivative in as much detail as possible.

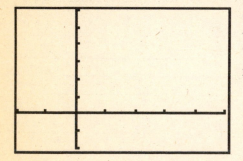

Change the viewing rectangle in order to get a very detailed view of both the function and its derivative *around* $x = 1$. The **TABLE** feature of your graphing calculator can also help you in analyzing the behaviors.

$[-5, 5]$ by $[-2, 6]$

Was it necessary to modify your original conjecture about the relationship between a function and its derivative? If so, describe these changes and the reasons for them in as much detail as possible.

64

Laboratory Exercise 4.2
The First Derivative Test and Intervals of Increasing and Decreasing

Name _____   Due Date _____

Enter each of the following functions along with its first derivative into your graphing calculator. You can find the derivative both analytically and graphically to make sure that you know the rules of differentiation.

(a)   As a test for yourself to see how well you understand the First Derivative Test, *'turn-off'* the derivative *of* the function, examine the graph of the function, and sketch how you believe the graph of the derivative should look on the first provided grid. (Consult your calculator manual to learn how to *"turn-off"* one or more of the *y-values*. On most calculators a function can be 'turned-off' by putting the calculator's cursor on the equal sign and hitting the **ENTER** key.) Identify those intervals of x's where the function is increasing and those where the function is decreasing.

(b)   Then do the opposite--- *'turn-off'* the function, examine the graph of the derivative, and sketch how the graph of the function should look. Use the second grid for your sketch.

(c)   Evaluate your sketches by *"turning-on"* both the function and its derivative and graphing them on the third grid provided.

(d)   Lastly, decide if the original function is even or odd. Then examine its derivative. Is the derivative even or odd? Can you make a conjecture and test your conjecture with some functions of your own?

1. $f(x) = x^2 + 3x - 4$

$[-8, 4]$ by $[-10, 20]$ $[-8, 4]$ by $[-10, 20]$ $[-8, 4]$ by $[-10, 20]$

Intervals of increasing_____

Intervals of decreasing_____

The function is even / odd.   The derivative is even / odd.

65

2. $f(x) = x^4 - 8x^2 + 16$

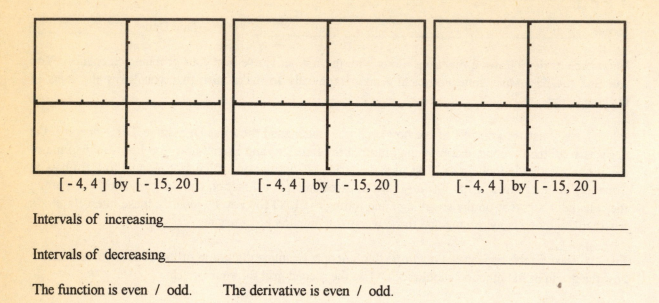

[- 4, 4] by [- 15, 20]   [- 4, 4] by [- 15, 20]   [- 4, 4] by [- 15, 20]

Intervals of increasing_____

Intervals of decreasing_____

The function is even / odd.   The derivative is even / odd.

3. $f(x) = \dfrac{x}{x^2 + 2}$

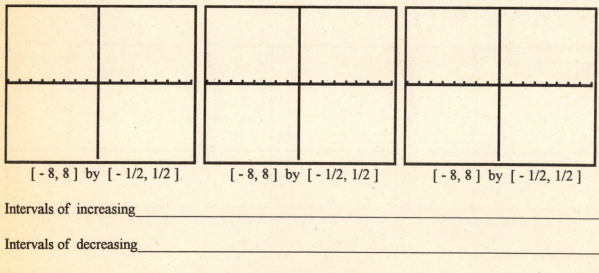

[- 8, 8] by [- 1/2, 1/2]   [- 8, 8] by [- 1/2, 1/2]   [- 8, 8] by [- 1/2, 1/2]

Intervals of increasing_____

Intervals of decreasing_____

The function is even / odd.   The derivative is even / odd.

4. $f(x) = (x-3)^2 (x+2)^3$

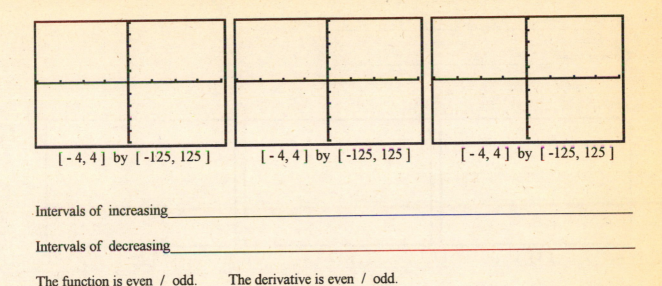

[-4, 4] by [-125, 125]   [-4, 4] by [-125, 125]   [-4, 4] by [-125, 125]

Intervals of increasing_____

Intervals of decreasing_____

The function is even / odd.   The derivative is even / odd.

5.   Now test your conjectures with two examples of your own. Give the function, its derivative, the viewing rectangle that you selected for your graphing calculator, and state whether the function and derivative are even or odd.

f(x) = _____

Derivative = _____

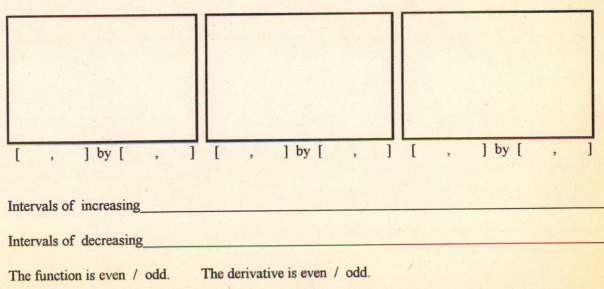

[,   ] by [,   ]   [,   ] by [,   ]   [,   ] by [,   ]

Intervals of increasing_____

Intervals of decreasing_____

The function is even / odd.   The derivative is even / odd.

6.   Give the function, its derivative, the viewing rectangle that you selected for your graphing calculator, and state whether the function and derivative are even or odd.

f(x) = _____

Derivative = _____

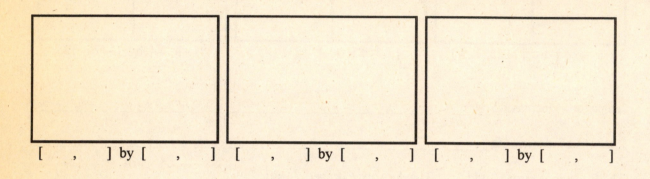

[,   ] by [,   ]   [,   ] by [,   ]   [,   ] by [,   ]

Intervals of increasing_____

Intervals of decreasing_____

The function is even / odd.   The derivative is even / odd.

Laboratory Exercise 4.3
The First Derivative Test and Maxima and Minima

Name _____   Due Date _____

Find all critical values of the following functions which also appeared in Laboratory Exercise 4.2. Classify the critical points as either stationary points or points of nondifferentiability. Explain your conclusions. Find the First Derivative either graphically by using the calculator's numerical derivative or analytically.

Then test yourself by *"turning-off"* the derivative of the function, examining the graph of the function, and deciding where the derivative should have zeros. (On most calculators a function can be 'turned-off' by putting the calculator's cursor on the equal sign and hitting the **ENTER** key.) Then do the opposite --- *"turn-off"* the function, examine the graph of the derivative, and decide where the function should have critical points and what type of critical point each should be.

1. $f(x) = x^2 + 3x - 4$

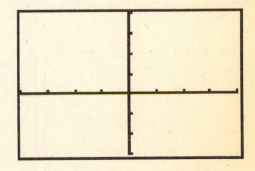

[- 8, 4] by [- 10, 20]

2. $f(x) = x^4 - 8x^2 + 16$

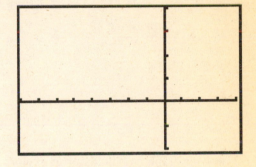

[- 4, 4] by [- 15, 20]

3. $f(x) = \dfrac{x}{x^2 + 2}$

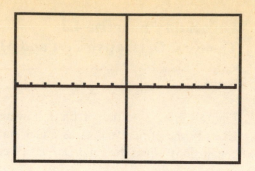

[- 8, 8] by [- 1/2, 1/2]

4. $f(x) = (x - 3)^2 (x + 2)^3$

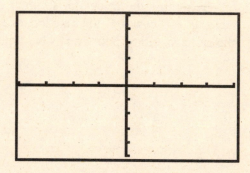

[- 4, 4] by [-125, 125]

Use your graphing calculator to explore various values for a in problems 5 and 6 given below and on the next page. Give the function and derivative graphs for three of the a's that you chose on the grid provided. Be sure to indicate the dimensions for each graph. Can you make a conjecture concerning the relative maxima and minima in each case?

5. $f(x) = x(2a - x)^{1/2}$

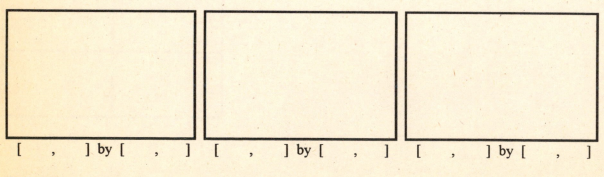

[,   ] by [,   ]   [,   ] by [,   ]   [,   ] by [,   ]

Conjecture:

70

6.　　　$f(x) = x(2ax - x^2)^{1/2}$

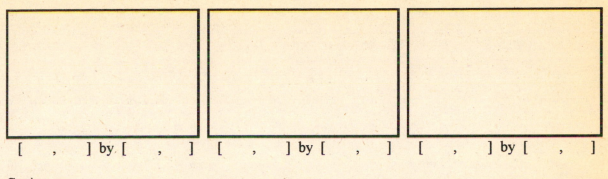

[　,　] by [　,　]　[　,　] by [　,　]　[　,　] by [　,　]

Conjecture:

7.　　　The following equation gives the bending moment M in a beam of length L and weight W. Find the algebraically least and greatest values of M given that $0 \le x \le L$.

$$M = -\frac{W}{2L}(x^2 - \frac{5Lx}{4} + \frac{L^2}{4})$$

Use your graphing calculator in any way that can prove helpful. Give a detailed description of how you came about your results.

Laboratory Exercise 4.4
The Second Derivative Test and Concavity

Name _____   Due Date _____

Graph each of the following functions and its second derivative on the provided grid. Determine where each function is concave up and where it is concave down. Explain how you came to your conclusions. Some graphing calculators have a **ROOT** finder feature. Use this feature to confirm your numerical results for the inflection points. Be careful as you graph each function and its second derivative. Would a change of scale for the viewing rectangle be necessary to examine the functions more carefully? Can you be sure that the vertical scale lets you see everything that you need to see in order to come to a correct conclusion(s)? If you do need to change the scale of the viewing rectangle, give the new dimensions and explain how you decided upon the new scale for the viewing rectangle.

1. $f(x) = x^2(x^2 - 10)$

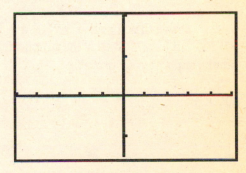

[- 4, 4] by [- 30, 10]

2. $f(x) = \dfrac{6}{x^2 + 3}$

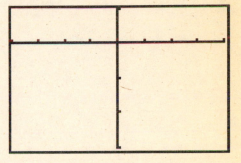

[- 5, 5] by [- 1.5, 2]

3. $f(x) = -4x^5 + 5x^3$

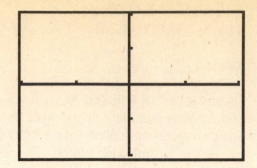

$[-2, 2]$ by $[-2, 2]$

4. $f(x) = \dfrac{x^2 + 3}{x^2 - 4}$

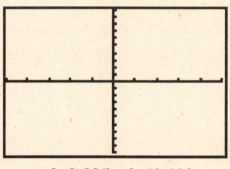

$[-5, 5]$ by $[-10, 10]$

5.   Prove algebraically and graphically that the parabola $x^2 = 2py$ and the hyperbola $xy = a^2$ have no points of inflection. Explain how you came to your conclusions.

6.   Prove that for all x, $x^2 - x + 1 > 0$. Given that this is true, how would you expect the graph of $f(x)$ to look? Sketch the graph of $f(x)$ and its second derivative on the provided grid. Do the graphs support your proof?

PROOF:

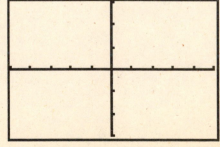

$[-5, 5]$ by $[-3, 3]$

74

Laboratory Exercise 4.5
Curve Sketching

Name _____   Due Date _____

In Questions 1 and 2 you are given the graph of a function *f(x)*. Identify those points where the derivative is zero and the intervals where the derivative is positive and negative. Then *sketch* the derivative on the same grid as the function. State whether the function is odd or even. Will the derivative be odd or even?

1.

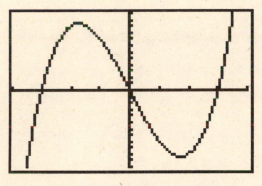

[- 4, 4] by [- 12, 12]

The function is odd / even.

The derivative is odd / even.

2.

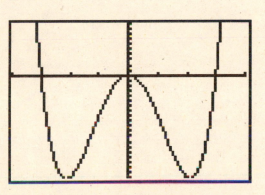

[- 4, 4] by [- 20, 10]

The function is odd / even.

The derivative is odd / even.

In Questions 3 and 4 you are given the graph of the derivative of a function *f(x)*. Identify the intervals where the original function is increasing or decreasing and where it has a maximum or minimum. Then *sketch* in the graph of a *possible* function on the same grid as the derivative.

3.

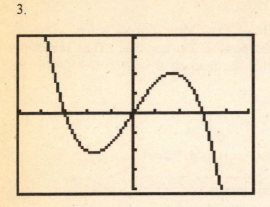

[- 5, 5] by [- 20, 20]

4.

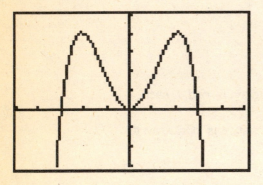

[- 5, 5] by [- 15, 25]

You have discovered in the previous exercises that the derivative of an odd function is even and the derivative of an even function is odd. Can the opposite be true also? That is to say, if the derivative is odd, will the function be even? If the derivative is even, will the function be odd? See if you can state conjectures about these questions by working through the following examples in 5 and 6.

5.   Graph the following three even functions and their derivatives on the provided grid.

```
Y₁■X²
Y₂■nDeriv(Y₁,X,X
)
Y₃■X²+2
Y₄■nDeriv(Y₃,X,X
)
Y₅■X²+6
Y₆■nDeriv(Y₅,X,X
```

```
WINDOW FORMAT
Xmin=-5
Xmax=5
Xscl=1
Ymin=-5
Ymax=10
Yscl=1
```

How many graphs did you actually sketch on the grid?_____

Why did this happen? _____

_____

The functions were even. Is/Are the derivative(s) odd or even?_____

Test this further with an example of your own involving even functions and their derivatives. In the space below give the functions you use and explain what happens with the derivatives.

6.   Now graph the following three functions and their derivatives on the provided grid.

```
Y₁■X
Y₂■nDeriv(Y₁,X,X
)
Y₃■X+3
Y₄■nDeriv(Y₃,X,X
)
Y₅■X-4
Y₆■nDeriv(Y₅,X,X
```

```
WINDOW FORMAT
Xmin=-4
Xmax=4
Xscl=1
Ymin=-5
Ymax=5
Yscl=1
```

How many graphs did you actually sketch on the grid?_____

State whether each function is odd or even. _____

Is/Are the derivative(s) odd or even? _____

Test this further with an example of your own involving functions that have even derivatives. In the space below give the functions you use and explain what happens with the derivatives.

Can you make a conjecture about the original function if you know its derivative is odd?

Can you make a conjecture about the original function if you know its derivative is even?

In problem 7 which is continued on the following pages match the function graph found in the left column with the correct derivative graph found in the right column.

7.   A) 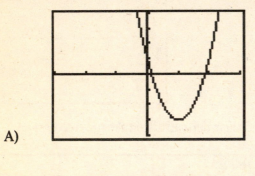   1)

      B) 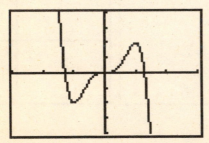   2)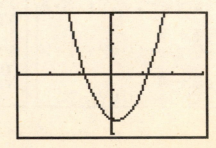

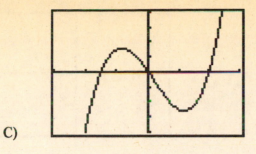 C)

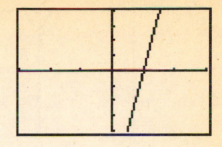 3)

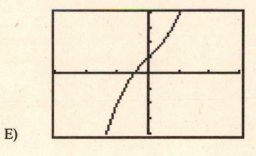 D)

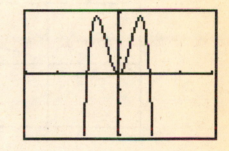 4)

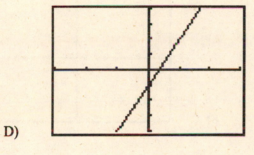 E)

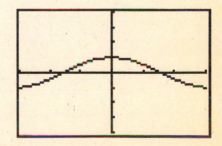 5)

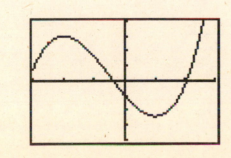 F)

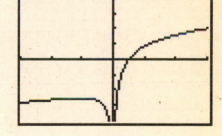 6)

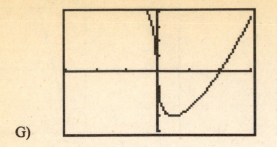

G)

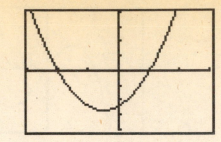

7)

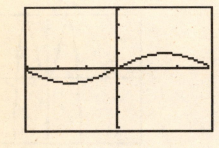

H)

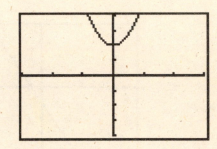

8)

The derivative of Graph A is Graph _____ .

The derivative of Graph B is Graph _____ .

The derivative of Graph C is Graph _____ .

The derivative of Graph D is Graph _____ .

The derivative of Graph E is Graph _____ .

The derivative of Graph F is Graph _____ .

The derivative of Graph G is Graph _____ .

The derivative of Graph H is Graph _____ .

Laboratory Exercise 4.6
Volume of a Cone

Name _____   Due Date _____

A cone is to be constructed from a circular piece of material that is 18 inches in diameter.

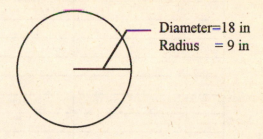

Diameter = 18 in
Radius   = 9 in

Cut along the radius and overlap to make it form a cone. Use your knowledge of derivatives to find the maximum volume of the cone. The equation for the volume of the cone with slant height of 9 inches is:

$$Volume = \frac{1}{3}\,\pi\,r^2\,\sqrt{9^2 - r^2}$$ where r is the radius of the *floor* of the cone.

Graph the function (r will become x on the calculator) in several different viewing rectangles and then decide upon one that you believe gives the best representation of the problem situation. Explain your reasoning. Be very careful with your decision. It is possible for the graph of the function to give misleading information if you do not take into consideration the pixel arrangement of your particular graphing calculator!!

Use **TRACE** and **ZOOM** to approximate the x-value that gives the maximum volume for the cone. If your calculator has a **TABLE** feature, use it to support your graphical approximation of the maximum. Use **TABLE SET** with a small ΔTbl to get an even better approximation of the maximum. Record your observations in the table below.

Radius (x) of Floor	Volume of Cone

Use the **MAXIMUM** feature of your calculator to find the maximum volume from your graph. Record your answer.

Now graph the derivative of the volume function on the same grid as the volume function. Use the **ROOT** finder on your calculator to find the root of the derivative function. Sketch your results below labeling the range and domain, the maximum of the volume function, and the root of the derivative function on your graph. What conclusions can you make from your findings?

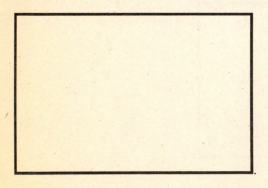

[,   ] by [,   ]

Laboratory Exercise 4.7
Newton's Method

Name _____   Due Date _____

Recall that the formula for **Newton's Method** is given by:

$$x_{n+1} = x_n - \frac{f(x_n)}{f'(x_n)} \; , \; for \; n = 1,2,3,4......$$

A simple way to use your calculator to perform Newton's Method is illustrated in Problem 1. Use this method to find the root(s) for each of the functions given in this exercise.

1.   Graph the function $f(x) = x^3 + 2x + 1$ on the grid below. Examine your graph to obtain initial guesses for any zeros of the function.

$y_1 = x^3 + 2x + 1$

$y_2 = nderiv(Y_1, x, x)$

[-2, 2] by [-5, 5]

Use the illustrated Newton's Method shown below to approximate any real zeros with an error of less than 0.0001. Now go to your **HOME SCREEN** and store your *initial guess for the zero*, say, -0.4, into x. Since Y1 is the original function and Y2 is the derivative of that function, you can now evaluate the zero of function following the above formula for Newton's Method by repeatedly hitting the **ENTER** key until your result starts to repeat itself or the accuracy is at the level you desire. Study the steps on the **HOME SCREEN** shown below.

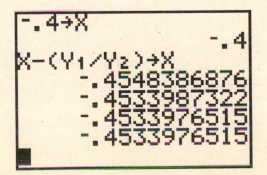

```
-.4→X
                -.4
X-(Y₁/Y₂)→X
        -.4548386876
        -.4533987322
        -.4533976515
        -.4533976515
■
```

2.   Graph the function $f(x) = 3x^3 - 5x + 1$ on the grid below. Examine your graph to obtain initial guesses for any zeros of the function. Then use the illustrated Newton's Method to obtain any real zeros with an error of less than 0.0001.

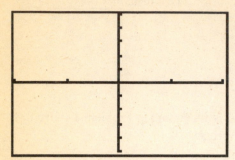

Solutions using Newton's Method:_____

[- 2, 2] by [- 5, 5]

3.   Graph the function $f(x) = x - \tan\left(x - \dfrac{\pi}{3}\right)$, $0 \le x \le 3$ on the grid below. Examine your

graph to obtain initial guesses for any zeros of the function. Then use the illustrated Newton's Method to obtain any real zeros with an error of less than 0.0001.

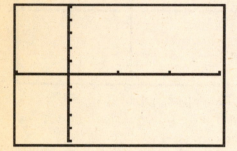

Solutions using Newton's Method:_____

[- 1, 3] by [- 5, 5]

4.   Graph the function $f(x) = 4 - 3x^3 - x^4$ on the grid below. Examine your graph to obtain initial guesses for any zeros of the function. Then use the illustrated Newton's Method to obtain any real zeros with an error of less than 0.0001.

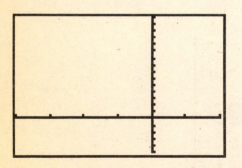

[- 4, 2] by [- 5, 15]

84

Chapter 5

Integration

Laboratory Exercise 5.1
Antiderivatives: The Indefinite Integral

Name _____   Due Date _____

Recall that a function F is the **antiderivative** of a function f if the derivative of F is equal to the function f for all x, in a given interval.

The process of finding antiderivatives is called **antidifferentiation or integration**.

MAKING A CONJECTURE: Use your graphing calculator to investigate each of the following functions and the derivative of each function. Graph each of these derivatives on the grid provided below and make a conjecture about functions that differ by a constant.

$$y_1 = x^3 \qquad\qquad y_2 = x^3 - 6 \qquad\qquad y_3 = x^3 + 5$$
$$y_4 = y_1{'} \qquad\qquad y_5 = y_2{'} \qquad\qquad y_6 = y_3{'}$$

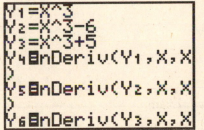

If you have a TI-82 graphing calculator, your screen should look like the one on the left. Notice that y_1, y_2, and y_3 are 'turned-off' in order to investigate only the graphs of the three derivatives.

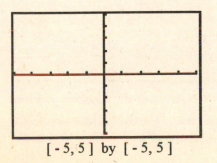

$[-5, 5]$ by $[-5, 5]$

Find each of the following general antiderivatives. Check your result graphically by graphing the function f and the derivative of F. Give the dimensions of your viewing rectangle when necessary.

1. $f(x) = 3x^3 - 4$

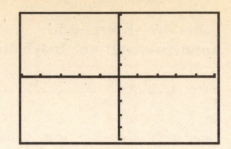

$[-5, 5]$ by $[-5, 5]$

2. $f(x) = x^4 - 2x^3 + 7$

3. $f(x) = 5\sin x + 3\cos x$

4. $f(x) = \dfrac{x^5 + 3x^2 - 2}{x^4}$

What you should have discovered is that $F'(x) = f(x)$. The antiderivatives of the function f differ only by a constant on any interval. Another way of saying this is that on any particular interval of x's the graphs of the antiderivatives of f are curves that are vertical translations of each other. You might even say that these vertical translations make up a 'family' of curves.

Laboratory Exercise 5.2
Area and Riemann Sums as a Function of the Number of Subintervals

Name _____   Due Date _____

Find the approximate area under the curve in each of the following problems over the interval [A, B] with the number of subintervals, n, being 5, 10, 20, and 50. Calculate the area in three ways:

- the left end point of each subinterval
- the right end point of cach subinterval
- the midpoint of each subinterval

An easy way to accomplish all of the arithmetic is to write a program for your graphing calculator. A sample program for the TI-82 referred to as the Rectangular Approximation Method (RAM) is given in the screen below. In the program N is the number of subintervals, L refers to left point rectangles, M refers to midpoint rectangles, and R refers to right point rectangles. Enter the equation into the Y_1. (*Reprinted with permission of Ray Barton and Addison-Wesley Publishing Company, Reading MA*)

```
PROGRAM:RAM
:Disp "A="
:Input A
:Disp "B="
:Input B
:Disp "NO.SUBINT
="
:Input N
```

```
PROGRAM:RAM
:(B-A)/N→H
:H/2→D
:A→X
:0→L
:0→M
:0→R
:Lbl 1
```

```
PROGRAM:RAM
:Y₁+L→L
:X+D→X
:Y₁+M→M
:X+D→X
:Y₁+R→R
:If B-X>.001
:Goto 1
```

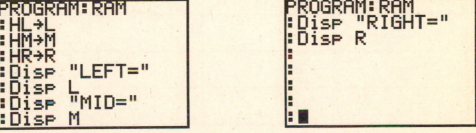

```
PROGRAM:RAM
:HL→L
:HM→M
:HR→R
:Disp "LEFT="
:Disp L
:Disp "MID="
:Disp M
```

```
PROGRAM:RAM
:Disp "RIGHT="
:Disp R



:
```

1. $f(x) = 4x - 2; \quad A = 1, B = 4$

n	5	10	20	50
LEFT				
MID				
RIGHT				

2. $f(x) = 4 - x^3 \quad A = -1, B = 1$

n	5	10	20	50
LEFT				
MID				
RIGHT				

3. $f(x) = x^3 - 4x \quad A = -2, B = 0$

n	5	10	20	50
LEFT				
MID				
RIGHT				

88

4. $f(x) = x^3 - 4x$ $A = 0, B = 2$

n	5	10	20	50
LEFT				
MID				
RIGHT				

What relationship do you see between Problems 3 and 4? What do you anticipate the answer to be for Problem 5 given below?

5. $f(x) = x^3 - 4x$ $A = -2, B = 2$

n	5	10	20	50
LEFT				
MID				
RIGHT				

Did you conjecture the correct result for #5? If not, what was your misconception?

Laboratory Exercise 5.3
The Fundamental Theorem of Calculus

Name _____   Due Date _____

In each of the following problems, evaluate the integral *without* using your graphing calculator. Then check your result by graphing the integral *and* your result on the *same* screen. What do you anticipate seeing in your viewing rectangle? An example of the input for the TI-82 graphing calculator is given below. Notice the use of a *dummy variable, A,* in the second equation.

EXAMPLE: $\displaystyle\int_{-a}^{x} (x^2 + 3x - 7)\, dx$

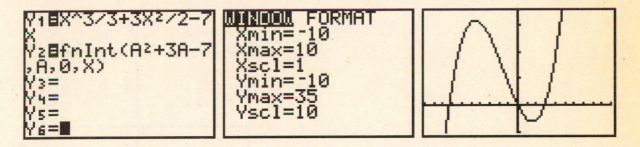

Use the **TRACE** feature of your graphing calculator to move between the two functions in order to compare their functional values for the x-values.

Now enter the original function in the third equation position and the derivative of the second equation into the fourth position. *'Turn-off'* the first and second functions. What do you anticipate when you **GRAPH** these two functions? Use the **TRACE** feature of your graphing calculator to move between the two functions in order to compare their functional values for the x-values.

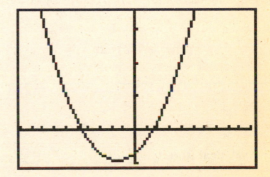

Now complete the following problems. After making observations of your own from these problems, you should be able to state Fundamental Theorem of Calculus in your words.

1. $\displaystyle\int \frac{2}{x^2+2}\ dx$

[,   ] by [,   ]

2. $\displaystyle\int \sin^2\theta\ d\theta$

Use the Fundamental Theorem of Calculus to evaluate the following integrals without using your graphing calculator. Then confirm your results by evaluating the integral with the graphing calculator.

3. $\displaystyle\int_{-\pi/2}^{\pi/2} (\cos^2 x + 2)\ dx$

Analytical Result_____ Calculator Result_____

4. $\displaystyle\int_{3}^{7} (2x^{-3} + 4x^{-2} - 3x^{-1}\ )\ dx$

Analytical Result_____ Calculator Result_____

5. $\displaystyle\int_{-\pi/2}^{\pi/2} (\cos\ x - \sin x)\ dx$

Analytical Result_____ Calculator Result_____

Now find the integral $\displaystyle\int_{-\pi/2}^{\pi/2} \cos\ x\ dx$ both analytically and with your calculator. What do you notice about the two answers?

Why do you think this happens? It might prove helpful if you graph the original function and sketch it on the grid provided below.

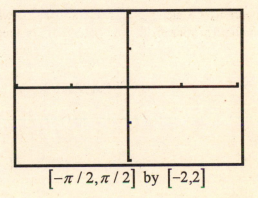

$\left[-\pi\,/\,2, \pi\,/\,2\right]$ by $\left[-2, 2\right]$

The Difference Rule for Definite Integrals tells us: $\displaystyle\int_a^b [f(x) - g(x)]dx = \int_a^b f(x)dx - \int_a^b g(x)dx$.

What result(s) do you get for the original problem when you work it both analytically and with the calculator using the Difference Rule? In other words, find

$$\int_{-\pi/2}^{\pi/2} [\cos x - \sin x]dx = \int_{-\pi/2}^{\pi/2} \cos x dx - \int_{-\pi/2}^{\pi/2} \sin x dx .$$

What results do you get for $\displaystyle\int_{-\pi/2}^{\pi/2} \cos x dx$? _____

What results do you get for $2 \displaystyle\int_{0}^{\pi/2} \cos x dx$? _____

Why do you get these answers? Explain your conclusions both in writing and with a graph.

$[-\pi/2, \pi/2]$ by $[-2,2]$

6. $\displaystyle\int_2^5 (4x^{3/2} + 3) \ dx$

Analytical Result_____   Calculator Result_____

7. $\displaystyle\int_{-3}^2 (8x^{2/3} - 5x + 4x^{-1}) \ dx$

Analytical Result_____   Calculator Result_____

94

Laboratory Exercise 5.4
Integration by Substitution

Name _____   Due Date _____

(a)   Evaluate the integral using the rules of integration you have learned. Check your
       results by differentiation.

(b)   Graph the integral and your analytical result on the same viewing rectangle. What
       do you expect to see?

On the first two problems remember to use a *dummy variable* as explained in Laboratory Exercise 5.3.

THOUGHT QUESTION: What would you expect to see in your viewing rectangle if you
graphed the integrand of the integral and *the derivative of the integral* on the same viewing rectangle?

1. $\displaystyle\int \frac{dx}{\sqrt{3x+7}}$

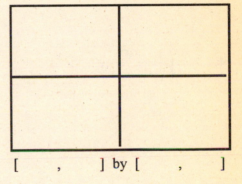

[,   ] by [,   ]

2. $\displaystyle\int \sqrt{5x-6}\ dx$

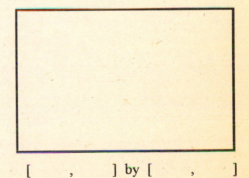

[,   ] by [,   ]

Evaluate the following definite integrals both analytically and by using the **HOME SCREEN** of your graphing calculator. If your calculator has the **CALC--->INTEGRAL** feature, verify your analytical result from the **GRAPH SCREEN** also. Show your results on the grid provided. Remember to set your viewing rectangle to a 'friendly' screen in order to be able to use your **TRACE** correctly for the integral lower and upper bounds!

3. $\displaystyle\int_{2}^{5} \sqrt{x^2 - 2}\ 2x\ dx$

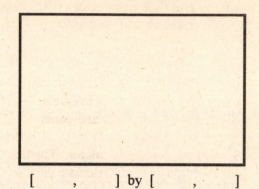

[,   ] by [,   ]

4. $\displaystyle\int_{3}^{4} \frac{x\,dx}{(x^2 - 4)^{3/2}}$

[,   ] by [,   ]

96

Chapter 6

Applications of the Definite Integral

Laboratory Exercise 6.1
Area Between Two Curves

Name _____   Due Date _____

Graph the two curves given in each problem on your graphing calculator and sketch your results in the provided grid. Find the points of intersection analytically and confirm with the **INTERSECT** utility of your calculator. Label the intersection points on your graph. Find the area between the two curves using your knowledge of integrals. Shade in the region of the graph for which you have found the area. Show all of your pencil-and-paper work.

Confirm with your calculator. This can be accomplished using the **2nd CALC #7** feature on the TI-82 graphing calculator. Calculate the integral of the first equation using the points of intersection that you found and then calculate the integral of the second function. Subtract the two results to compare your results with those of the calculator. This can also be accomplished on the **HOME SCREEN**. The calculator steps for the TI-82 are shown below for Problem 1.

1. $\qquad y = 5 - 5x^2 \qquad$ and $\qquad y = 2 - 2x^2$

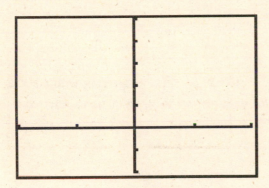

$$[-2, 2] \text{ by } [-2, 5]$$

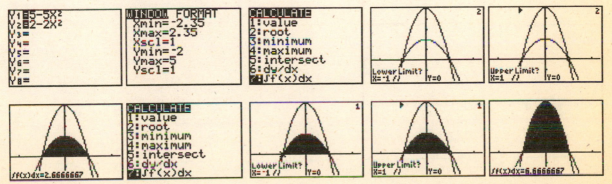

Utilizing the **HOME SCREEN** can be accomplished in the following manner:

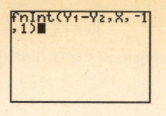

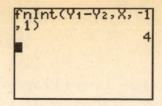

2. $y = x^2$ and $y = x + 2$

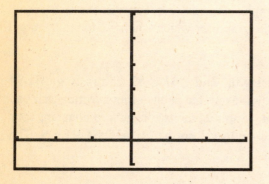

[- 3, 3] by [- 1, 5]

3.   Find the area between $y^2 = x + 2$ and $y = x$,   first integrating with respect to x (sometimes referred to as using vertical strips) and then with respect to y (sometimes referred to as using horizontal strips).

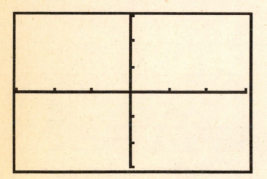

[- 3, 3] by [- 3, 3]

98

4. $y = x(4 - x)$ and $y = x$ from $x = 0$ to $x = 4$. Be careful! It might be wise to examine the graph very carefully first and then find the area analytically and confirm with your calculator second!

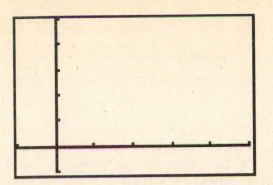

$[-1, 5]$ by $[-1, 5]$

5.   Find the area between the graph of $y = x^3 - 12x$ and the tangent line drawn at the maximum point on the curve. Sketch the graph and this tangent on the grid and shade in the area.

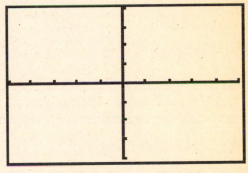

$[-5, 5]$ by $[-20, 20]$

Now find the area between the graph of $y = x^3 - 12x$ and the tangent line drawn at the minimum point on the curve. Sketch the graph and this tangent on the grid below and shade in the area.

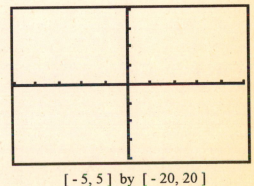

$[-5, 5]$ by $[-20, 20]$

99

6. Let $y = \dfrac{Px^2}{R^2}$ where P and R are positive constants. Use your graphing calculator to explore the graphs of y for various values of P and R along with the graph of $y = P$. For each of your cases, find the area in the first quadrant between the curve $y = \dfrac{Px^2}{R^2}$, the y-axis, and the line $y = P$ using your graphing calculator. Record your findings in the table below.

P	R	Area

For your different values of P and R, do you see a pattern for the values of the integrals (areas between the curves)? What is it?

Confirm your results be first solving the problem by integrating with respect to y and then with respect to x. Do your integration results support the conjecture that you stated above?

Laboratory Exercise 6.2
Volume

Name _____   Due Date _____

When working with Volumes of Solids of Revolution, it is best to sketch a picture of the figure first. Use your graphing calculator to help you do this and sketch in your results in the grid provided including the solid and the disc or shell/washer, depending on the method that you are using. Keep in mind that the region can be revolved about any line, not just the x- or y-axis. Show all of your work in the space provided. Use your calculator to confirm your results whenever possible.

1.   Use the disc method to find the volume of the resulting solid when the region bounded by the x-axis, $y = x^2$, and $x = 2$ is revolved about the x-axis. **NOTE**: Some graphing calculators have the capability to **DRAW VERTICAL LINE**s. The following screens show how this is accomplished on the TI-82 graphing calculator using the **DRAW** key. Once you have positioned the vertical line as near as possible to $x = 2$, press the **ENTER** key to 'secure' your vertical line. Then press the **GRAPH** key to take you out of the **DRAW** utility. Of course, you could always draw the line $x = 2$ by hand.

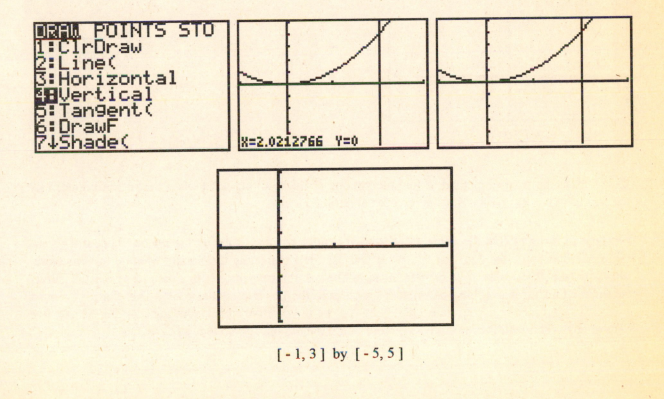

[- 1, 3] by [- 5, 5]

2.   Use the disc method to find the volume of the resulting solid when the region bounded by the y-axis, $y = x^2$, and $y = 2$ is revolved about the y-axis.

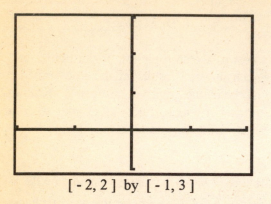

[-2, 2] by [-1, 3]

3.   Use the washer method to find the volume of the solid in Problem 2 on the previous page. How should your answers compare?

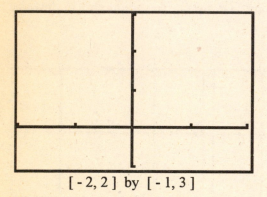

[-2, 2] by [-1, 3]

4.   Use the washer method to find the volume of the resulting solid when the region bounded by the y-axis and $x = y(y - 4)$ is revolved about the x-axis.

NOTE: In order to find the volume of the solid resulting from revolving this region about the x-axis, you must first solve the equation for y. If you use the Completing the Square Method in your work, you will take the square root of both sides of the resulting equation. In order to graph the curve, remember that most graphing calculators graph only the positive square root. Therefore, you must input two equations into your calculator. Let Y_1 be the positive square root and Y_2 be the negative square root. This will then give you a picture of a parabola opening to the right.

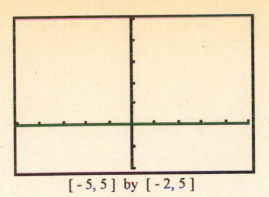

[-5, 5] by [-2, 5]

Suppose you wanted only an approximation of the volume of this solid? How could you go about this? SUGGESTION: Draw a triangle inside the region of your curve in the second quadrant using the two y-intercepts and the vertex of the parabola as the three vertices of the triangle. Draw this on the grid below.

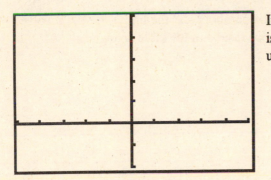

If you were to rotate this region about the x-axis, how close is this result to the result you got above when you were using the parabola??

[-5, 5] by [-2, 5]

5.   Use your graphing calculator to help you find the volume of the solid generated by revolving the area between the curve $y = (\sqrt{a} - \sqrt{x})^2$ and the coordinate axes about the y-axis by examining several values of a. Record your findings for your selected values of a in the table below. Sketch the curves for three values of a in the grid to the right.

a	Area of solid

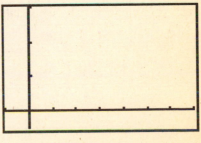

[-1, 7] by [-1/2, 3]

How does the value of a influence the resulting graphs? How does the value of a influence the bounded area?

After these explorations, can you find the volume of this solid for any value of a?

Confirm your volume conjecture by integration.

6.   Now recall the curve $y = \dfrac{Px^2}{R^2}$ that we examined in Problem 6 in the previous exercise. This time examine the volume generated when the area between the parabola and the line $y = P$ is revolved about the y-axis for various values of P and R. Record your results in the table below.

P	R	Volume of the Solid

Can you conjecture about the volume in general terms of P and R?

Check your conjecture by using integration.

Laboratory Exercise 6.3
Arc Length and Surface Area

Name _____   Due Date _____

1.   Find the arc length of the curve $y = x^{2/3}$ from $x = 1$ to $x = 8$ analytically. Support with your graphing calculator using the **FNINT** utility. Be careful that you enter the equation into the calculator correctly by referring to the calculator manual for the correct syntax. Sketch the curve on the grid below. Then find the length of the line segment joining the points (1, 1) to (8, 4). How does the length of the line segment compare with the arc length?

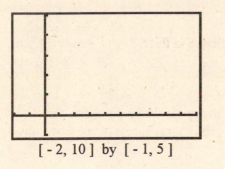

[- 2, 10] by [- 1, 5]

2.   Find the arc length of the curve $y = \dfrac{x^3}{6} + \dfrac{1}{2x}$ from $x = 1$ to $x = 4$ analytically. Support with your graphing calculator. Sketch the curve on the grid below. Then find the length of the line segment joining the points (1, 2/3) to (4, 259/24). How does the length of the line segment compare with the arc length?

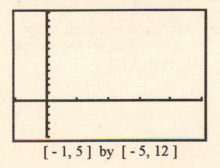

[- 1, 5] by [- 5, 12]

What if you were to take another point between these two points on the curve, say the midpoint, and join these points forming two smaller line segments. How does the sum of these two smaller line segments compare with the arc length?

105

3.   Find the arc length of the curve $x = \dfrac{y^3}{3} + \dfrac{1}{4y}$ from $y = 1$ to $y = 2$ analytically. Support with your graphing calculator. Sketch the curve on the provided grid. Then find the length of the line segment joining the points (7/12 , 1) to (67/24 , 2). How does the length of the line segment compare with the arc length?

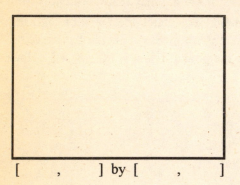

[,   ] by [,   ]

What if you were to take two more equally spaced points between the original two points on the curve and then join these points forming three smaller line segments. How does the sum of these three smaller line segments compare with the arc length?

4.   Find the arc length of the curve $x = y^{3/2}$ from $y = 1$ to $y = 4$ analytically. Support with your graphing calculator. Sketch the curve on the grid below. Then find the length of the line segment joining the points (1, 1) to (8 , 4). How does the length of the line segment compare with the arc length? Do you see any similarity between this curve and the one given in Problem 1?

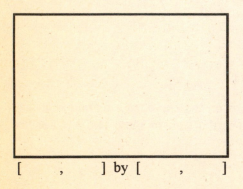

[,   ] by [,   ]

106

What if you were to take three points between the original two points on the curve and join these points forming four smaller line segments. How does the sum of these four smaller line segments compare with the arc length?

Can you make a conjecture about arc length compared to the sum of smaller and smaller line segments?

CONJECTURE:

5.   Find the area of the surface generated by revolving the curve $y = \dfrac{x^4}{4} + \dfrac{1}{8x^2}$ from $x = 2$ to $x = 4$. Find a way to use your graphing calculator to confirm your result.

6.   Find the area of the surface generated by revolving the curve $x = \dfrac{y^2}{8}$ from $y = 2$ to $y = 4$. Find a way to use your graphing calculator to confirm your result.

Laboratory Exercise 6.4
Applications of Integration to Rectilinear Motion

Name _____   Due Date _____

1.   A truck is traveling with a constant acceleration of 8 ft/sec^2. Its initial velocity is 12 ft/sec. How far will the truck travel in 20 seconds? Show all of your paper-and-pencil work in the space below.

Distance = _____

You can simulate the truck's path by using **PARAMETRIC MODE** on your graphing calculator. For instance, for this problem the sequence of steps would be as follows on the TI-82 graphing calculator where $Y_{1T} = 3$ is arbitrarily chosen.

```
      Sci Eng      |WINDOW FORMAT  |WINDOW FORMAT  |X₁ᴛ◼4T²+12T |                     1
Float 0123456789   |Tmin=0         |↑Tstep=.1      |Y₁ᴛ◼3       |
Radian Degree      |Tmax=20        | Xmin=-100     |X₂ᴛ=        |
Func Par Pol Seq   |Tstep=.1       | Xmax=1900     |Y₂ᴛ=        |
Connected Dot      |Xmin=-100      | Xscl=100      |X₃ᴛ=        |
Sequential Simul   |Xmax=1900      | Ymin=-1       |Y₃ᴛ=        |
FullScreen Split   |Xscl=100       | Ymax=5        |X₄ᴛ=        | T=20
                   |↓Ymin=-1       | Yscl=1◼       |Y₄ᴛ=◼       | X=1840        Y=3
```

QUESTION: What is the velocity at the end of the 20 seconds? Is this reasonable?
Work in groups to answer these questions. What parameters of the problem situation could be changed to make the problem reasonable? What would you change? Rewrite this problem so that it is reasonable situation.

2.   The truck driver is traveling at a speed of 72 ft/sec when he suddenly slams on the brakes. This causes the truck to decelerate at a constant rate of 12 ft/sec^2. How far will the truck travel before coming to a complete stop?

Distance = _____

Simulate this using the **PARAMETRIC EQUATION** capability of your graphing calculator. Indicate below the equations you used for the simulation.

$X_{1T} =$ _____ $Y_{1T} =$ _____

Is this problem reasonable the way it is given?   Explain your reasoning.

3.   A basketball player throws the ball at an angle of $40°$ when the ball is 6 feet above the floor. The basketball reaches a maximum height of 28 feet. How far will the ball travel horizontally before it hits the floor?

Can you simulate this motion on your calculator?

Recalling your work with arc length in the previous exercise, can your find the *actual* distance that the basketball traveled (this would be the arc length)? Indicate all of your mathematical steps below. Can you simulate this on your graphing calculator?

4.   A particle travels on a coordinate line at time t with a velocity of $v(t) = t^2 - t$ m/sec. Give the integral for the displacement of the particle during the time interval $[0, 3]$. Then evaluate the integral.

Find the integral that represents the distance traveled by the particle during the same time interval. Evaluate the integral.

Chapter 7

Logarithmic and Exponential Functions

Laboratory Exercise 7.1
Inverse Functions

Name _____ Due Date _____

Most graphing calculators have a **DRAW INVERSE** feature that enables you to draw the inverse of a function. Care must be taken, however, when using the **DRAW INVERSE** feature of many graphing calculators. Try the following example. The screens shown are for the TI-82 graphing calculator. Consult your calculator's manual to see how to graph the inverse of a function on your particular calculator.

Example: Graph the function $y = x^3 + 2x - 1$ and the line $y = x$ on the same screen. (It is easier to see the inverse of a function if you choose dimensions for your viewing rectangle that are fairly *square*.) Use the Horizontal Line Test to check if the cubic function has an inverse. If it does, **DRAW THE INVERSE** on the same viewing rectangle.

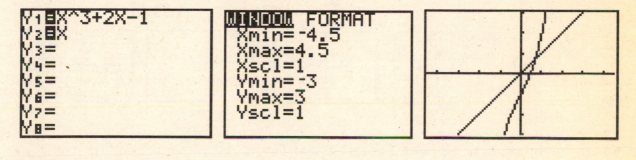

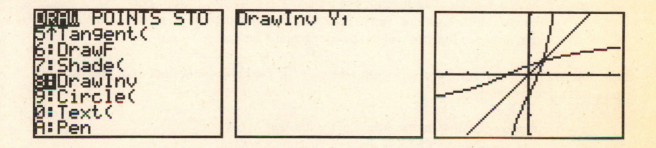

You can see that to draw the inverse of the original function you reflect the original function about the line $y = x$. The two graphs are *mirror images* of each other. Furthermore, by investigating the graphs of the two functions, you can see that the point (1, 2) is on the original graph $y = x^3 + 2x - 1$ and the point (2, 1) appears to be on the inverse graph. (Unfortunately, on the TI-82 you cannot

TRACE on the inverse function when using **DRAW INVERSE**.) In general, if the point (a, b) is a point in the function f then the point (b, a) is a point in the inverse of f denoted by f^{-1}.

WORD OF CAUTION: With some graphing calculators, care must be taken when using the **DRAW INVERSE** utility. As an example, consider the even function $y = x^2$. It is an even function and fails the Horizontal Line Test. However, the TI-82 will draw in inverse of this even function *even though it does not exist!* The calculator draws the image of the parabola reflected about the line $y = x$! You know this cannot be possible, because $y = x^2$ is an even function! Yet, the calculator does, indeed, draw it even though an inverse does not exist! Another example is the function $y = \tan x$. The second row of calculator screen below shows you how the calculator **DRAW**s the **INVERSE** of this function.

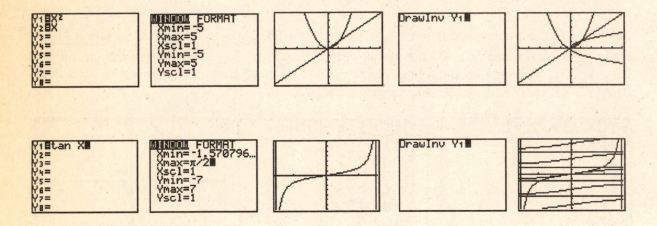

Now try some examples on your own. **For each of the following:**

    a.   Sketch a graph of the function and the line $y = x$ on the provided grid.
    b.   Use the Horizontal Line Test to determine if the function has an inverse. (Remember that any even function will fail the Horizontal Line Test.)

If the function has an inverse, proceed with steps c - e. If the inverse does not exist, explain how you came to that conclusion.

    c.   Find an equation for the inverse of the function.
    d.   Use **DRAW INVERSE** to draw the function's inverse.
    e.   Find some points (a, b) on the original function and see if the points (b, a) are on the graph of the inverse.

1. $y = \dfrac{3}{x-3}$

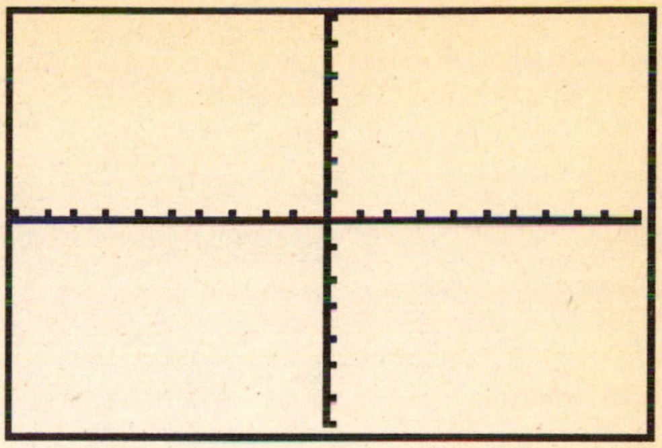

$f^{-1} = $ _____

$[-10, 10]$ by $[-7, 7]$

Points on f _____

Points on f^{-1} _____

2. $y = \dfrac{2}{x}$

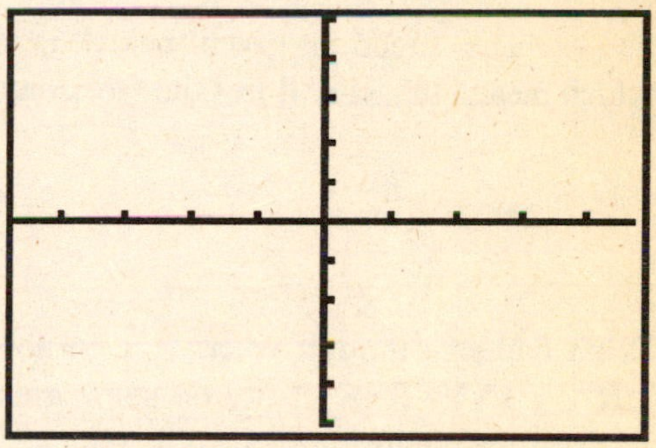

$f^{-1} = $ _____

Points on f _____

$[-4.5, 4.5]$ by $[-5, 5]$

Points on f^{-1} _____

3. $y = \sqrt{2x + 5}$

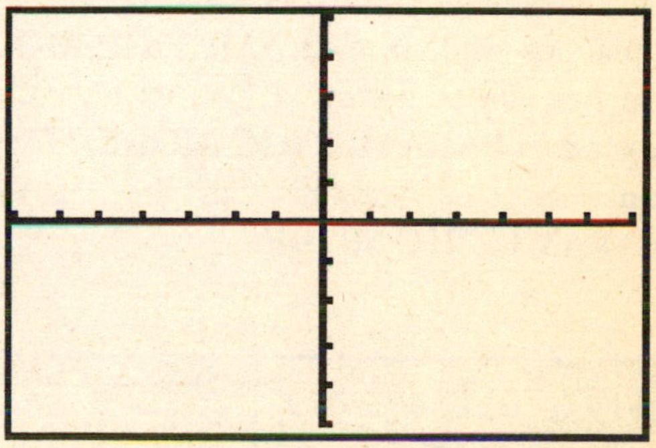

$f^{-1} = $ _____

Points on f _____

$[-7, 7]$ by $[-5, 5]$

Points on f^{-1} _____

113

4. $y = x^3$

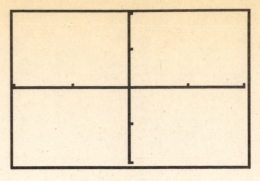

[-2, 2] by [-2, 2]

$f^{-1} =$ _____

Points on f _____

Points on f^{-1} _____

5.   The function given below is an even function which means that it will fail the Horizontal Line Test!

$$y = \frac{6}{x^2 + 1}$$

What happens though when you go ahead and apply the **DRAW INVERSE** utility on some graphing calculators? Does the calculator appear to be telling you that an inverse for y does exist? This is definitely one of the *Pitfalls of Technology* that you should be aware of!

[-6, 6] by [- 6, 6]

It is also possible to graph a function and its inverse using parametric equations. The **DRAW** utility does not allow you to **TRACE** on the function or its inverse. However, you can **TRACE** on both equations while in **PARAMETRIC MODE** making it possible to **TRACE** to the point $(a, f(a))$ on the function and $(f(a), a)$ on its inverse. *As* an example, graph the equation in Exercise 1 and its inverse in **PARAMETRIC MODE**. The following steps are for the TI-82 graphing calculator. You must also be careful with the **DRAW INVERSE** utility of some graphing calculators in **PARAMETRIC MODE**.

```
Normal Sci Eng
Float 0123456789
Radian Degree
Func Par Pol Seq
Connected Dot
Sequential Simul
FullScreen Split
```

```
WINDOW FORMAT
Tmin=0
Tmax=10
Tstep=.1
Xmin=-9.4
Xmax=9.4
Xscl=1
↓Ymin=-10
```

```
WINDOW FORMAT
↑Tstep=.1
Xmin=-9.4
Xmax=9.4
Xscl=1
Ymin=-10
Ymax=10
Yscl=1
```

```
X1т=T
Y1т=3/(T-3)
X2т=Y1т
Y2т=X1т
X3т=
Y3т=
X4т=
Y4т=
```

114

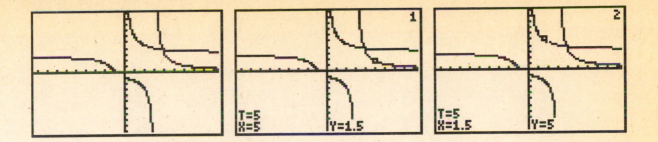

Now graph the following functions and their inverses in **PARAMETRIC MODE**. In each case, give the parametric equations that you used.

6. $y = \sqrt{2x + 5}$

$X_{1T} = $ _____

$Y_{1T} = $ _____

$X_{2T} = $ _____

$Y_{2T} = $ _____

[0, 10] by [- 1, 10]

7. $y = x^3$

$X_{1T} = $ _____

$Y_{1T} = $ _____

$X_{2T} = $ _____

$Y_{2T} = $ _____

[- 2, 2] by [- 2, 2]
$-10 \le T \le 10$

115

Laboratory Exercise 7.2
Graphs of $y = \ln x$ **and** $y = e^x$

Name _____   Due Date _____

1.   Plot each of the following functions on the grid provided below.

  $y = \ln ax$, for $a = 1, 2, 3,$ and 4 .

What conjecture can you make about this family of
curves?

Without using your graphing calculator, how do you
anticipate the graph of $y = \ln 5x$ to look? Sketch in
your answer and support with your graphing calculator.

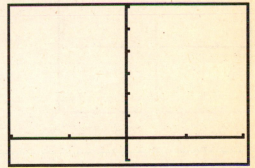

[- 2, 2] by [-3, 6]

2.   Plot each of the following functions on the grid provided below.

  $y = e^{nx}$, for $n = 1, 2, 3,$ and 4 .

What conjecture can you make about this family of
curves?

Without using your graphing calculator, how do you
anticipate the graph of $y = e^{5x}$ to look? Sketch in
your answer and support with your graphing calculator.

[- 2, 2] by [- 1, 6]

117

3.   Sketch the following three equations on the grid below: $y = x$ $y = \ln x$ $y = e^x$. Then put all that you have observed in Problems 1 and 2 together with these sketches and make a conjecture about these two families of functions.

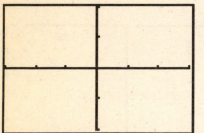

[-3, 3] by [-2, 2]

Conjecture:

Use your **DRAW INVERSE** feature to first draw the inverse of $y = \ln x$. What do you notice?

Now **DRAW** the **INVERSE** of $y = e^x$. What do you notice?

Graph the function $y = \ln x$ and its inverse in **PARAMETRIC MODE** using the procedure that you learned in Exercise 7.1. Sketch the graphs on the grid provided below. Give the parametric equations that you used.

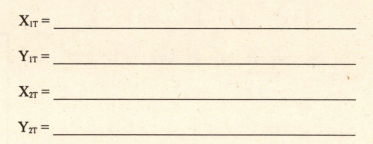

$X_{1T} =$ _____

$Y_{1T} =$ _____

$X_{2T} =$ _____

$Y_{2T} =$ _____

[-3, 3] by [-2, 2]
Time: [0, 3, 0.1]

You can also repeat this exercise with $y = e^x$. What graph would you expect to obtain?

4. Use your graphing calculator to explore the following functions on the same viewing rectangle.

$$y_1 = e^x \qquad y_2 = 1+x \qquad y_3 = \left(1+\frac{x}{2}\right)^2 \qquad y_4 = \left(1+\frac{x}{4}\right)^4 \qquad y_5 = \left(1+\frac{x}{8}\right)^8$$

Sketch each function on the grid to the right. What appears to be happening to the graphs as the denominator of x and the exponent of the function get larger?

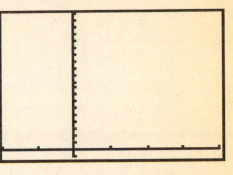

[- 2, 4] by [- 1, 20]

Now plot $y_1 = e^x$ and $y_2 = \left(1+\frac{x}{100}\right)^{100}$ on the same grid to the right. Then make a conjecture about the $\lim\limits_{x \to \infty} e^x$.

CONJECTURE:

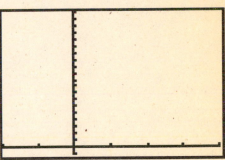

[- 2, 4] by [- 1, 20]

5. Graph $y = \ln(1+e^x)$ and $y = x$ on the same grid. What do you see happening as x gets larger and larger?

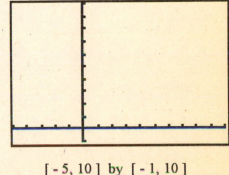

[- 5, 10] by [- 1, 10]

119

Laboratory Exercise 7.3
Developing the Properties of the Natural Logs Graphically
Solving Equations Involving Logs

Name _____   Due Date _____

Part I: Developing the Log Identity Rules Graphically

1.   Graph each of the following functions on the same graph. How many graphs do you see when you graph the four functions? Use **TRACE** to help you identify those functions that are equal or work in groups with each student graphing a different equation on his/her calculator. Then compare graph results to see which graphs are the same. (The screens shown here are for the TI-82 graphing calculator. The input for other graphing calculators should look similar.)

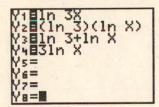

```
Y1=ln 3X
Y2=(ln 3)(ln X)
Y3=ln 3+ln X
Y4=3ln X
Y5=
Y6=
Y7=
Y8=■
```

```
WINDOW FORMAT
 Xmin=-1
 Xmax=3
 Xscl=1
 Ymin=-3
 Ymax=3
 Yscl=1
```

Which equations are equal? _____

Repeat the exploration as follows:

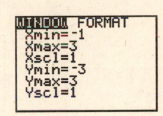

```
Y1=ln .5X
Y2=(ln .5)(ln X)

Y3=ln .5+ln X
Y4=.5ln X
Y5=
Y6=
Y7=
```

```
WINDOW FORMAT
 Xmin=-1
 Xmax=3
 Xscl=1
 Ymin=-3
 Ymax=3
 Yscl=1
```

Which equations are equal? _____

Try some other combinations of equations on your own. Do you obtain the same results? Does it make a difference if the 0.5 is changed to - 0.5? Can you make a conjecture about a rule of logs?

CONJECTURE:

2.   Graph each of the following functions on the same graph. How many graphs do you see when you graph the four functions? Use **TRACE** to help you identify those functions that are equal or work in groups with each student graphing a different equation on his/her calculator. Then compare graph results to see which graphs are the same.

```
Y₁█ln (X/3)
Y₂█(ln 3)(ln X)
Y₃█ln 3+ln X
Y₄█ln 3-ln X
Y₅█ln X-ln 3
Y₆=
Y₇=
Y₈=■
```

```
WINDOW FORMAT
 Xmin=-1
 Xmax=3
 Xscl=1
 Ymin=-3
 Ymax=3
 Yscl=1
```

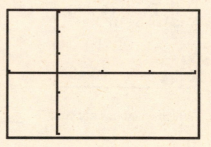

Which equations are equal? _____

Repeat the exploration as follows:

```
Y₁█ln (X/7)
Y₂█(ln 7)(ln X)
Y₃█ln 7+ln X
Y₄█ln 7-ln X
Y₅█ln X-ln 7
Y₆=
Y₇=
Y₈=
```

```
WINDOW FORMAT
 Xmin=-1
 Xmax=3
 Xscl=1
 Ymin=-3
 Ymax=3
 Yscl=1
```

Which equations are equal? _____

Try some other combinations of equations on your own. Do you obtain the same results? Can you make a conjecture about a rule of logs?

CONJECTURE:

3. Graph each of the following functions on the same graph. How many graphs do you see when you graph the four functions? Use **TRACE** to help you identify those functions that are equal or work in groups with each student graphing a different equation on his/her calculator. Then compare graph results to see which graphs are the same.

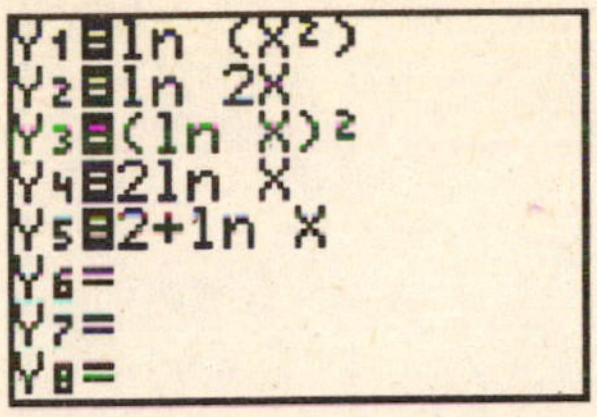

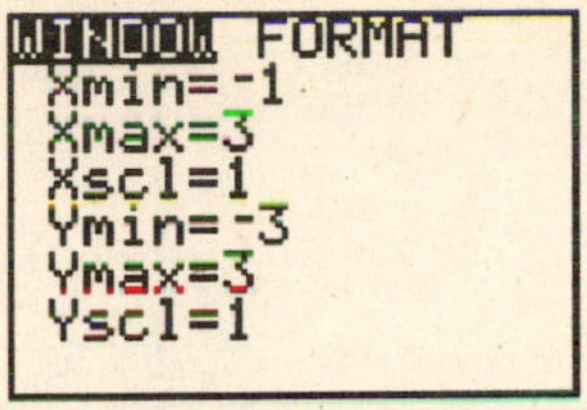

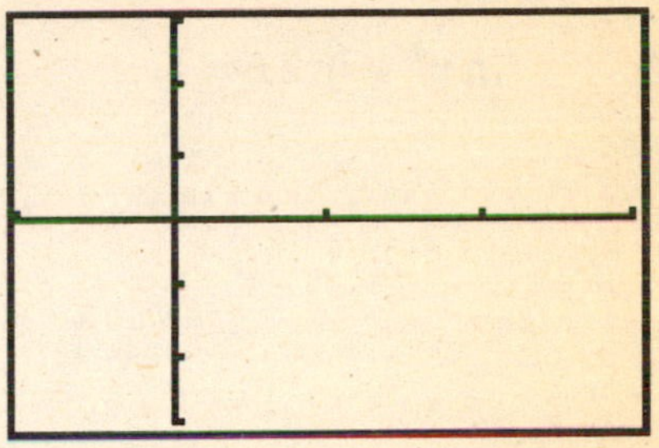

Does it make a difference whether the exponents are positive or negative? ____________________

Which equations are equal for positive exponents? ____________________

Repeat the exploration as follows:

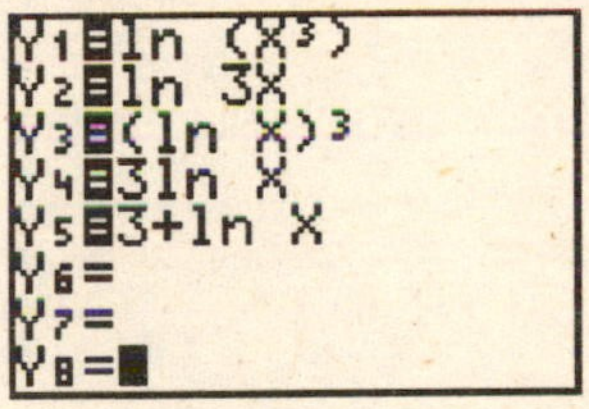

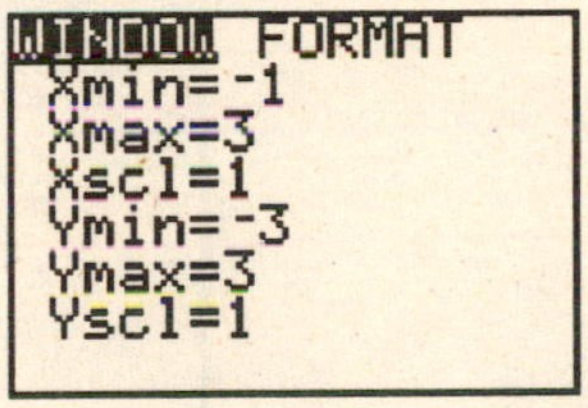

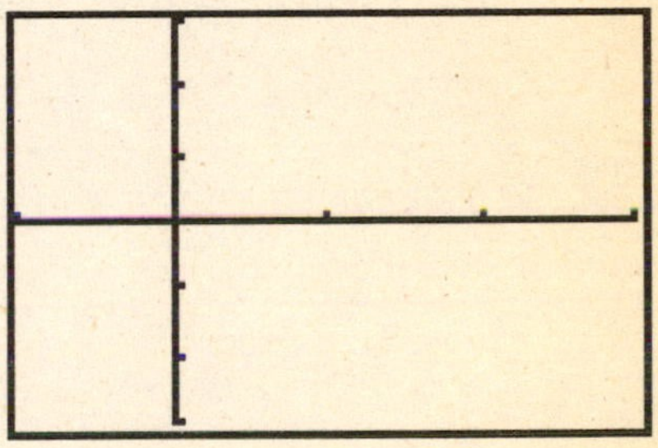

Which equations are equal for positive exponents? ____________________

Try some other combinations of equations on your own. Do you obtain the same results? Can you make a conjecture about a rule of logs?

CONJECTURE:

Part II: Solving Equations Involving Logs

Solve each of the following equations using your graphing calculator and the multi-graph method of solving equations. Graph your results. Verify your results algebraically. Show all of your work.

4. $\ln x^2 - \ln 2x = 1$

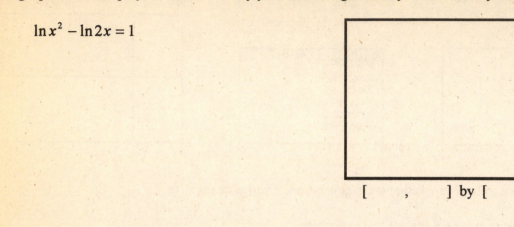

[,   ] by [,   ]

5. $\ln x = 4\ln 2 - 2\ln 3 - \ln 5$

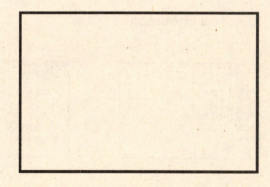

[,   ] by [,   ]

Laboratory Exercise 7.4
Derivatives and Integrals of Logarithmic and Exponential Functions

Name _____   Due Date _____

Find the derivative of each of the following functions analytically. Support your answer by graphing both your answer and **nDERIV** on the same grid. What do you expect to see?

1. $y = \ln \dfrac{x^2(x+1)}{(x-3)^3}$

This derivative is quite messy analytically! How can you use the rules of natural logs to simplify your work?

[,   ] by [,   ]

2. $y = \ln(x + \sqrt{x^2 - 1})$

[,   ] by [,   ]

3. $y = xe^x$

[,   ] by [,   ]

4. $y = \dfrac{1 - e^x}{x^2}$

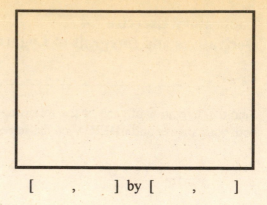

[,   ] by [,   ]

Find the integral of each of the following functions analytically. Support your answer by graphing both the original function (without the integral) and the **nDERIV** of your answer on the same grid. If you have obtained the correct integral, how many graphs would you expect to see?

5. $\displaystyle\int \dfrac{x}{x^2 + 1}\,dx$

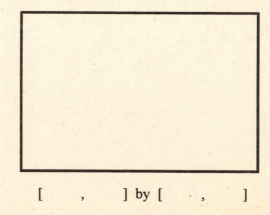

[,   ] by [,   ]

6. $y = \displaystyle\int x e^{x^2}\,dx$

[,   ] by [,   ]

7. Evaluate $\int_0^2 xe^x\, dx$ analytically. Then use your **RAM PROGRAM** to evaluate the definite integral for different values of n. How do the two compare?

If you are using the TI-82 graphing calculator, you can also use the **2nd CALC #7** feature of the calculator. What answer do you get for the integral when you use this feature?

8. The integral $\int_0^1 e^{\frac{-x^2}{2}}\, dx$ is seen often in the study of probability. However, you cannot to this integral analytically. You can, however, use graphing calculator technology to evaluate this integral.

Use your calculator to graph the function $y = e^{\frac{-x^2}{2}}$. Use the **RAM PROGRAM** to find the value of the integral for different values of n. What answer do you get?

If your graphing calculator has the capability to do definite integrals, use it to evaluate the integral. What result do you obtain?

How do the two answers compare?

Laboratory Exercise 7.5
Hyperbolic Functions----Identities, Derivatives, and Integrals

Name _____   Due Date _____

The first two exercises are good opportunities for group work. Let one student graph the sum of the two given functions and another student graph the hyperbolic function. Then examine *both* graphs at the same time to see how they compare. If you are working singularly, then use the TRACE feature of your graphing calculator to help you decide if the sum of the two functions and the hyperbolic function are the same.

1.   Graph $y_1 = \dfrac{e^x}{2}$ and $y_2 = \dfrac{-e^{-x}}{2}$
on the grid to the right.

Then graph $y_3 = y_1 + y_2$ on the same grid.

Let $y_4 = \sinh x$ and graph it on the same grid. How do the third and fourth graphs compare?

$[-3, 3]$ by $[-3, 3]$

Is this what you were expecting? Explain your reasoning.

2.   Graph $y_1 = \dfrac{e^x}{2}$ and $y_2 = \dfrac{e^{-x}}{2}$
on the grid to the right.

Then graph $y_3 = y_1 + y_2$ on the same grid.

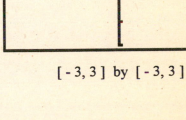

Let $y_4 = \cosh x$ and graph it on the same grid. How do the third and fourth graphs compare?

$[-3, 3]$ by $[-3, 3]$

Was this what you were expecting? Explain your reasoning.

3.   Graph the following equation as two graphs to decide if it is an identity. If it is an identity, give a proof using your knowledge of hyperbolic functions. *If it is not an identity*, find all points that make the equation true.

$$1 - \tanh^2 x = \sec h^2 x$$

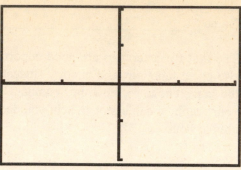

$$[-2, 2] \text{ by } [-2, 2]$$

4.   Graph the following equation as three graphs to decide if it is an identity. Which equations, if any, are identities? If there is an identity, give a proof using your knowledge of hyperbolic functions. *If it is not an identity*, find all points that make the equation true.

$$\cosh^2 x + \sinh^2 x = 2\sinh^2 x + 1 = 2\cosh^2 x$$

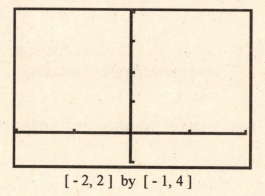

$$[-2, 2] \text{ by } [-1, 4]$$

5.   Graph the following equation as two graphs to decide if it is an identity. If it is an identity, give a proof using your knowledge of hyperbolic functions. *If it is not an identity*, find all points that make the equation true within the viewing rectangle interval.

$$\cosh^2 x = \frac{1}{2}\left(1 + \cosh 2x\right)$$

$$[-1, 1] \text{ by } [-1, 3]$$

6.   Find the derivative of $y = f(x) = x \sinh x$. Use your graphing calculator to support your answer by graphing the derivative of y and your answer on the same grid. If your result is correct, what do you expect to see?

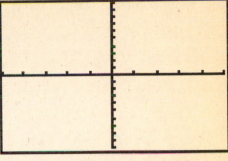

[- 5, 5] by [- 10, 10]

7.   Find the derivative of $y = f(x) = \cosh(4x + 1)$. Use your graphing calculator to support your answer by graphing the derivative of y and your answer on the same grid. If your result is correct, what do you expect to see?

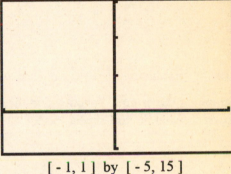

[- 1, 1] by [- 5, 15]

8.   Find the derivative of $y = f(x) = \cos \sinh x$. Use your graphing calculator to support your answer by graphing the derivative of y and your answer on the same grid. If your result is correct, what do you expect to see?

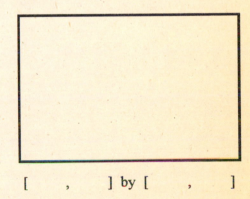

[,   ] by [,   ]

131

9.   Evaluate $\int \tanh^2 x\,dx$. Use your graphing calculator to support your answer by graphing the original function (without the integral sign) and the **NDERIVIV** of your answer on the same grid. If your result is correct, what do you expect to see?

[,   ] by [,   ]

10.   Evaluate $\int \sinh^2 x \cosh x\,dx$. Use your graphing calculator to support your answer by graphing the original function and the **NDERIVIV** of your answer on the same grid. If your result is correct, what do you expect to see?

[,   ] by [,   ]

Chapter 8

Inverse Trigonometric Functions and Inverse Hyperbolic Functions

Laboratory Exercise 8.1
Inverse Trigonometric Functions

Name _____   Due Date _____

1.   On the grid given below are the graphs of $Y_1 = \sin x$ and $Y_2 = x$. On the same grid graph the inverse of $\sin x$, $\sin^{-1} x$. Recall that the inverse of the sine function exists for $-\pi/2 \leq x \leq \pi/2$. Discuss the range of this inverse function. Support your graph by entering $\sin^{-1} x$ into the Y_3 position and examining the resulting graph. Using the **ZOOM #7 TRIG** utility on the TI-82 calculator would give a graph like the following:

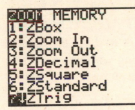 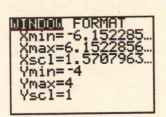 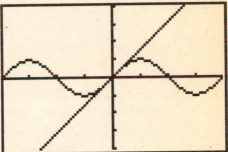

Domain of x _____

Range of y _____

QUESTION: If your graphing calculator has a **DRAW INVERSE** utility and you use it to draw the inverse of $\sin x$ do you obtain the same graph? Sketch this result on the grid below along with $\sin x$. Is the graph the same as the one you obtained above for $\sin^{-1} x$? Why or why not? The screen shown here are for the TI-82 graphing calculator.

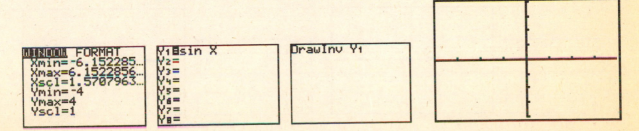

133

2.   On the grid given below are the graphs of $Y_1 = \cos x$ and $Y_2 = x$. On the same grid graph the inverse of $\cos x$, $\cos^{-1} x$. Discuss the domain and range of this inverse function.

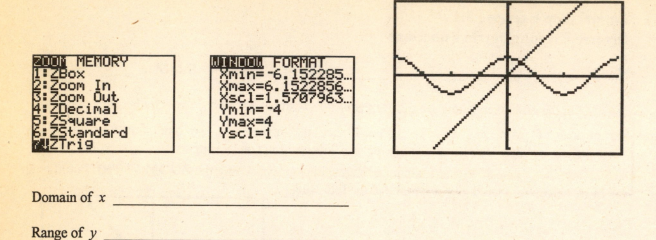

Domain of x _____

Range of y _____

QUESTION: If you use the **DRAW INVERSE** utility of your graphing calculator to draw the inverse of $\cos x$ do you obtain the same graph? Sketch this result on the grid below along with $\cos x$. Is the graph the same as the one you obtained above for $\cos^{-1} x$? Why or why not?

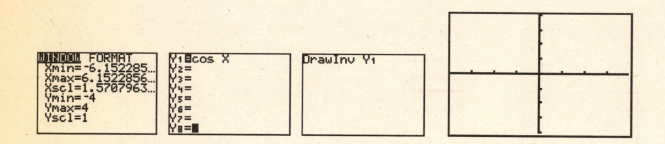

3.   Use your graphing calculator to help you find the domain(s) of each of the basic relationships $\sin^{-1}(\sin x) = x$ and $\sin(\sin^{-1} x) = x$. Sketch the graphs on the grids below indicating the domain for each relationship. Are they the same?

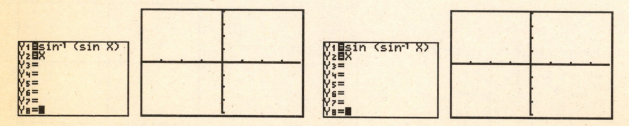

3.   Use your knowledge of trigonometric function and inverses of functions to *sketch* the graph of $y = 3\sin^{-1} 4x$ on the provided grid. To the right of the grid show all of your algebra work. Be sure to include the dimension of your graph. Use your graphing calculator **only to support** your sketch.

[,   ] by [,   ]

4.   Use your knowledge of trigonometric function and inverses of functions to *sketch* the graph of $y = 2\cos^{-1}(x/3)$ on the provided grid. To the right of the grid show all of your algebra work. Be sure to include the dimension of your graph. Use your graphing calculator **only to support** your sketch.

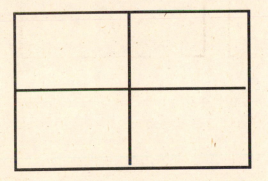

[,   ] by [,   ]

5.	For which values of x is $\tan^{-1}(\tan x) = x$? Use your graphing calculator to help you find this domain. Sketch the results on the provided grid. Confirm your graphical conclusion analytically.

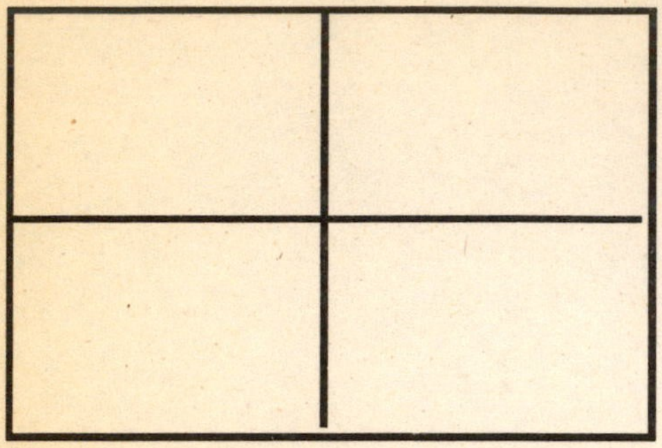

[,   ] by [,   ]

6.	For which values of x is $\tan(\tan x^{-1}) = x$? Use your graphing calculator to help you find this domain. Sketch the results on the provided grid. Confirm your graphical conclusion analytically.

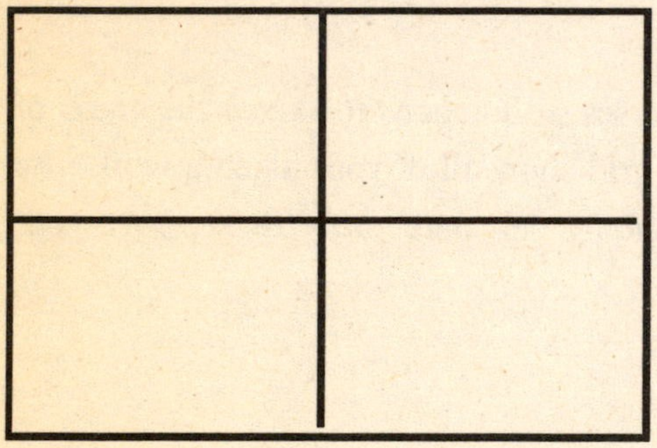

[,   ] by [,   ]

Laboratory Exercise 8.2
Derivatives and Integrals of Inverse Trigonometric Functions

Name _____   Due Date _____

1.   Find the derivative of $y = \sin^{-1} x^2$. Show all of your work below. Support your result with your graphing calculator by first graphing your result and then having the calculator graph the derivative of y. What do you expect to see when you use the **TRACE** utility on both graphs? Show the graphing results on the provided grid below.

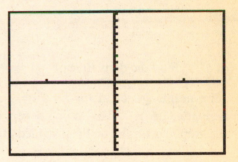

[- 1.5, 1.5] by [- 10, 10]

2.   Find the derivative of $y = \cos^{-1}(2x - 1)$. Show all of your work below. Support your result with your graphing calculator by first graphing your result and then having the calculator graph the derivative of y. What do you expect to see when you use the **TRACE** utility on both graphs? Show the graphing results on the provided grid below.

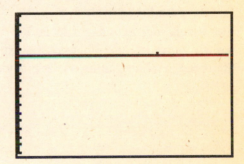

[0, 1.5] by [- 12, 5]

137

3.   Find the derivative of $y = \sqrt{1 - x^2} + \cos^{-1} x$. Show all of your work below. Support your result with your graphing calculator by first graphing your result and then having the calculator graph the derivative of y. What do you expect to see when you use the **TRACE** utility on both graphs? Show the graphing results on the provided grid below.

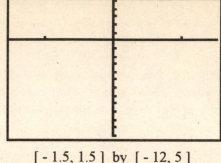

$[-1.5, 1.5]$ by $[-12, 5]$

4.   Find the derivative of $y = \dfrac{\sin^{-1} x}{x}$. Show all of your work below. Support your result with your graphing calculator by first graphing your result and then having the calculator graph the derivative of y. What do you expect to see when you use the **TRACE** utility on both graphs? Show the graphing results on the provided grid below.

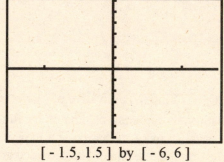

$[-1.5, 1.5]$ by $[-6, 6]$

5.   Find the derivative of $y = \tan^{-1} \sqrt{x^2 - 1}$. Show all of your work below. Support your result with your graphing calculator by first graphing your result and then having the calculator graph the derivative of y. What do you expect to see when you use the **TRACE** utility on both graphs? Show the graphing results on the provided grid below.

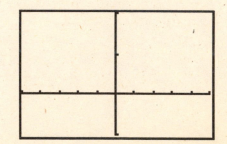

6.   It is known that for the functions f_1 and f_2 if there exists an interval [a, b] such that $f_1'(x) = f_2'(x)$ in [a, b], then there exists a number c such that $f_1(x) = f_2(x) + c$ in [a, b].

Show that $\tan^{-1} \dfrac{x+1}{x-1} + \tan^{-1} x = c$, where c is a constant, and then find c. Use your graphing calculator to help you find c graphically and confirm your findings analytically. Sketch the graphs of the two functions on the first grid and then sketch the graphs of the derivatives on the second grid. Indicate the interval(s) in which the derivatives are equal. If there is more than one interval, find the c for each interval.

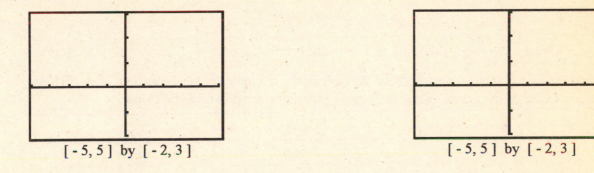

[- 5, 5] by [- 2, 3]   [- 5, 5] by [- 2, 3]

7.   Evaluate $\displaystyle\int \dfrac{2x\,dx}{5+x^2}$ without using your graphing calculator. Show all of your work below.

Support your result with your graphing calculator by first graphing your answer and then having the calculator graph the integral of y. You may need to refer back to Exercise 5.3 to review how to graph an integral using a *dummy variable*. An example of the input for the TI-82 graphing calculator is given in that exercise. What do you expect to see on the graph? Show the graphing results on the grid provided on the next page.

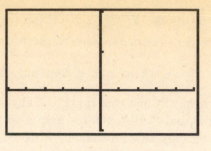

$[-5, 5]$ by $[-1, 2]$

QUESTION: Why do you obtain *two* graphs on your graphing calculator? How can you alter the integral you obtained so that when you graph it with the calculator's integral you will see only *one* graph?

8.   Evaluate $\int \dfrac{4dx}{3x - 2}$. Show all of your work below. Support your result with your graphing calculator by first graphing your result and then having the calculator graph the integral of y. What do you expect to see on the graph? Show the graphing results on the provided grid below.

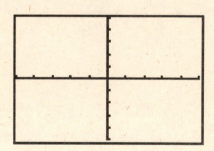

$[-5, 5]$ by $[-5, 5]$

Laboratory Exercise 8.3

Inverse Hyperbolic Functions and Derivatives and Integrals of Inverse Hyperbolic Functions

Name _____   Due Date _____

1.   Find the logarithmic equivalent of the inverse hyperbolic $\sinh^{-1} 3$. Show all of your work. Confirm your result on the **HOME SCREEN** of your graphing calculator by evaluating both $\sinh^{-1} 3$ and your answer.

2.   Find the logarithmic equivalent of the inverse hyperbolic $\tanh^{-1}(1/2)$. Show all of your work. Confirm your result on the **HOME SCREEN** of your graphing calculator by evaluating both $\tanh^{-1}(1/2)$ and your answer.

3.   Find the derivative of $y = \cosh^{-1}(3x + 2)$. Show all of your work. Support your result by graphing your answer and the derivative that your graphing calculator calculates. What do you expect to see?

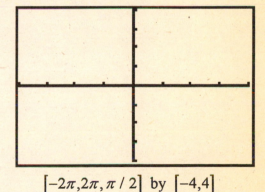

$\left[-2\pi, 2\pi, \pi / 2\right]$ by $\left[-4, 4\right]$

4.   Find the derivative of $y = (\sinh^{-1} x)^2$. Show all of your work. Support your result by graphing your answer and the derivative that your graphing calculator finds. What do you expect to see?

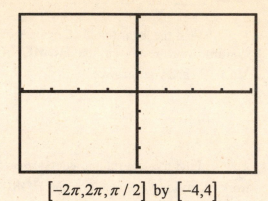

$[-2\pi, 2\pi, \pi/2]$ by $[-4,4]$

5.   Find the derivative of $y = x\sinh^{-1} x$. Show all of your work. Support your result by graphing your answer and the derivative that your graphing calculator finds. What do you expect to see?

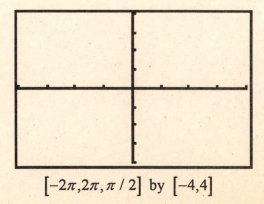

$[-2\pi, 2\pi, \pi/2]$ by $[-4,4]$

6. Evaluate $\displaystyle\int \frac{dx}{2x\sqrt{1+4x^2}}$. Show all of your work. Support your result by graphing your

answer and the integral that your graphing calculator finds (be sure to use a dummy variable). What do you expect to see? Note that the domain of the provided grid starts at 0.01 since $x \neq 0$.

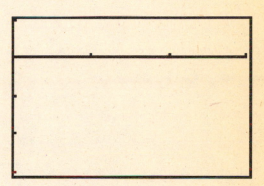

[0.01, 3] by [- 3, 1]

7. Evaluate $\displaystyle\int \frac{dx}{x\sqrt{1+x^8}}$. Show all of your work. Support your result by graphing your answer

and the integral that your graphing calculator finds. What do you expect to see?

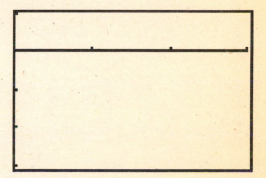

[0.01, 3] by [- 3, 1]

143

Chapter 9

Techniques of Integration

Laboratory Exercise 9.1
Integration by Parts

Name _____   Due Date _____

By this time you should be very comfortable with the rules of differentiation, the idea of an antiderivative, and the basic rules of integration. This chapter will continue with integration and various techniques of integration. Some of these techniques are rather lengthy pencil-and-paper computations. All the more reason to be confident in interpreting what it is exactly that your graphing calculator is telling you-----*and* to know <u>and</u> understand more than one way to support your analytical solution with the graphing calculator.

The *Integration by Parts Formula* is: $\int u\,dv = uv - \int v\,du$. There are, of course, different possibilities for the *u* and *dv* when applying the Integration by Parts Formula. This formula allows us to change certain integrals into forms that can be evaluated by methods that have already been developed.

As an example, consider the following integral that should be tackled using integration by parts.

Evaluate $\int_{-1}^{4} xe^{-x}\,dx$ analytically. Then support your result with your graphing calculator. How many ways can you think of to use the graphing calculator as a supporting device for your analytical answer?

Letting $u = x$ and $dv = e^{-x}$ would yield the analytical solution: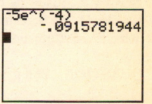
$(-xe^{-x} - e^{-x}) \left. \right|_{-1}^{4} = -5e^{-4}$

You can evaluate this using the **Home Screen** of your graphing calculator:

Now look at the problem graphically by following the steps given in the calculator screens below.

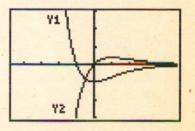

145

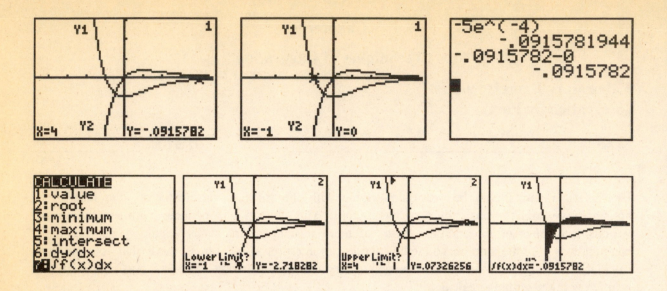

Explain in detail what is happening on the calculator screens. How many ways of supporting the analytical answer are shown?

Now try the following problems on your own. Show all of your work to obtain the analytical result. Use the **Home Screen** to confirm your answer and *also* use your graphing utility to reinforce your numerical confirmation. Sketch in the graphs on the provided grids.

QUESTION: Will it make a difference which term you select to be u?

1. $\displaystyle\int_{-1}^{1} x^3 e^{x^2}\, dx$ $u =$ $dv =$

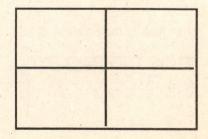

$uv - \displaystyle\int v\, du \;=$

146

2. $\int\limits_0^1 \ln x^2 \, dx$ $u =$ $dv =$

$uv - \int v \, du =$

3. $\int\limits_0^1 \ln 3x \, dx$ $u =$ $dv =$

$uv - \int v \, du =$

4. $\int\limits_0^\pi x^3 \sin 2x \, dx$ $u =$ $dv =$

$uv - \int v \, du =$

147

5. $\displaystyle\int_0^4 x \ln x \, dx$ $u =$ $dv =$

$uv - \int v \, du =$

6. $\displaystyle\int_{-\pi}^{\pi} x \cos x \, dx$ $u =$ $dv =$

$uv - \int v \, du =$

Now that you see that the integral is zero, can you think of an easier way to do this type of integral?

On your own find the integral of several other odd functions on an integral of $[-a, a]$. What is the result each time? Can you form a conjecture based upon your observations?

CONJECTURE:

7. $\displaystyle\int_{-1}^1 e^x \cos x \, dx$ $u =$ $dv =$

$uv - \int v \, du =$

148

Laboratory Exercise 9.2
Integrals of Trigonometric Functions

Name _____   Due Date _____

Evaluate each of the following integrals. Show all of your work. Use your graphing calculator to provide graphical support in the two ways and numerical support in the one way that were illustrated in Exercise 9.1. Since this is an indefinite integral you may want to review Exercise 5.3 on the use of *dummy variables*.

1. $\int \sin^3 x \, dx$

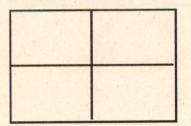

2. $\int \sin^2 x \cos x \, dx$

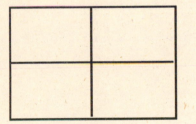

3. $\int \sin^5 x \, dx$

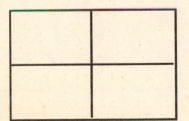

4. $\displaystyle\int_0^\pi \sin^2 x \cos^5 x \, dx$

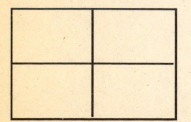

5. $\displaystyle\int \sec^3 x \tan^3 x \, dx$

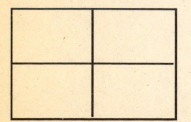

6. $\displaystyle\int \frac{\sin x}{3 + \cos x} \, dx$

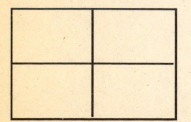

7. $\displaystyle\int \sec^2 x \tan 2x \, dx$

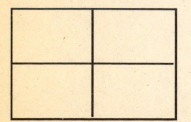

Laboratory Exercise 9.3
Trigonometric Substitutions

Name _____   Due Date _____

1.   Evaluate $\int \dfrac{x^3}{\sqrt{x^2+4}}\,dx$. Show all of your work. Indicate the trigonometric substitution that you use and the resulting dx. Use your graphing calculator to provide graphical support in two ways and numerical support in one way.

Substitution: $x =$ _____

Yielding $dx =$ _____

There are at least three substitutions that you can use to help you solve this integral. You used the trigonometric substitution above. Can you give two other substitutions that you could use that are not trigonometric?

2.   Evaluate $\int \dfrac{\sqrt{25-x^2}}{x^2}\,dx$. Show all of your work. Indicate the substitution that you use and the resulting dx. Use your graphing calculator to provide graphical support in two ways and numerical support in one way.

Substitution: $x =$ _____

Yielding $dx =$ _____

3. Evaluate $\int \dfrac{dx}{(16+x^2)^2}$. Show all of your work. Indicate the substitution that you use and the resulting dx. Use your graphing calculator to provide graphical support in two ways and numerical support in one way.

Substitution: $x =$ _____

Yielding $dx =$ _____

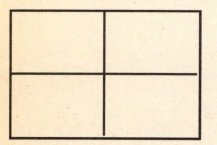

4. Evaluate $\int \dfrac{1}{x\sqrt{x^2+9}}\, dx$. Show all of your work. Indicate the substitution that you use and the resulting dx. Use your graphing calculator to provide graphical support in two ways and numerical support in one way.

Substitution: $x =$ _____

Yielding $dx =$ _____

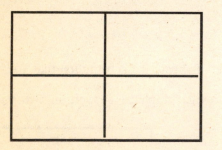

Laboratory Exercise 9.4
Integrating Rational Functions Using Partial Fraction Decomposition

Name _____   Due Date _____

Evaluate each of the following integrals by first expressing the fraction as a sum of partial fractions. Support your decomposition of the fraction by *graphing both* the original fraction and <u>your</u> decomposition result. What do you expect to see on your graph? Show all of your work and list the constants, A, B, C, and so on that you found and write out the new integral. Then evaluate the decomposed integral and support with your graphing calculator.

1.   Evaluate $\int \dfrac{dx}{(x+1)(x+2)(x+3)}$.

2.   Evaluate $\int \dfrac{dx}{x^2(x-3)^2}$.

3. Evaluate $\int \dfrac{\cos x \, dx}{\sin^2 x - 2 \sin x - 3}$.

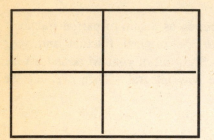

4. Evaluate $\int \dfrac{2x^2 + 1}{(x-2)^3} dx$.

5. Evaluate $\int \dfrac{(x^2 + 2x + 2) \, dx}{x^2 (x^2 + 2)^2}$.

154

Laboratory Exercise 9.5
Trapezoidal Rule and Simpson's Rule

Name _____   Due Date _____

The Trapezoidal Rule and Simpson's Rule are easy to program into a graphing calculator. Below are versions of these programs for the TI-82 graphing calculator. If you have a different calculator you may need to edit these programs to run on your calculator. Note that the function you want to take the integral of must be stored in *Y1*. *A* is the lower bound of the integral while *B* is the upper bound. (*Reprinted by permission of Ray Barton and Addison-Wesley Publishing Company, copyright 1994.*)

Program: Trapezoidal Rule

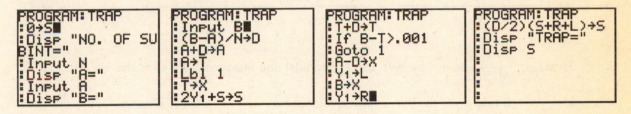

Program: Simpson's Rule

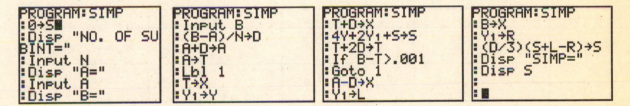

1. Evaluate $\int_{1}^{3}(x^3 - 4)dx$ by both the Trapezoidal and Simpson's Rule for the n's given in the table below. Which rule gives the best approximation for a given number of intervals? Then compare these results to those found using the RAM program from Exercise 5.2. Of the five results from the two programs, which rule gives the best approximation for a given number of intervals? Also find the *exact* answer using the Fundamental Theorem of Calculus. Show all of your work.

n	2	6	10	30
TRAP				
SIMP				

2. Evaluate $\int_0^1 \sqrt{x+1}\,dx$ by both the Trapezoidal and Simpson's Rule for the n's given in the table below. Which rule gives the best approximation for a given number of intervals? Then compare these results to those found using the RAM program from Exercise 5.2. Of the five results from the two programs, which rule gives the best approximation for a given number of intervals? Also find the *exact* answer using the Fundamental Theorem of Calculus. Show all of your work.

n	2	6	10	30
TRAP				
SIMP				

3. Evaluate $\int_0^6 \dfrac{dx}{\sqrt{x^2+1}}$ by both the Trapezoidal and Simpson's Rule for the n's given in the table below. Which rule gives the best approximation for a given number of intervals? Then compare these results to those found using the RAM program from Exercise 5.2. Of the five results from the two programs, which rule gives the best approximation for a given number of intervals? Also find the *exact* answer using the Fundamental Theorem of Calculus. Show all of your work.

n	2	6	10	30
TRAP				
SIMP				

4. Evaluate $\int_0^3 x\sqrt{4x+1}\,dx$ by both the Trapezoidal and Simpson's Rule for the n's given in the table below. Which rule gives the best approximation for a given number of intervals? Then compare these results to those found using the RAM program from Exercise 5.2. Of the five results from the two programs, which rule gives the best approximation for a given number of intervals? Also find the *exact* answer using the Fundamental Theorem of Calculus. Show all of your work.

n	2	6	10	30
TRAP				
SIMP				

5. Evaluate $\int_1^3 \sqrt{x^2 - x}\, dx$ by both the Trapezoidal and Simpson's Rule for the n's given in the table below. Which rule gives the best approximation for a given number of intervals? Then compare these results to those found using the RAM program from Exercise 5.2. Of the five results from the two programs, which rule gives the best approximation for a given number of intervals? Also find the *exact* answer using the Fundamental Theorem of Calculus. Show all of your work.

n	2	6	10	30
TRAP				
SIMP				

6. The integral $\int_1^2 e^{-x^2}\, dx$ cannot be done analytically. Evaluate it by both the Trapezoidal and Simpson's Rule for the n's given in the table below. Which rule gives the best approximation for a given number of intervals? Then compare these results to those found using the RAM program from Exercise 5.2. Of the five results from the two programs, which rule gives the best approximation for a given number of intervals? Also find the *exact* answer using the Fundamental Theorem of Calculus. Show all of your work.

n	2	6	10	30
TRAP				
SIMP				

7. Evaluate $\int_{-1}^1 3x - 2x^3 + \tan 2x\, dx$ by both the Trapezoidal and Simpson's Rule for the n's given in the table below. Which rule gives the best approximation for a given number of intervals? Then compare these results to those found using the RAM program from Exercise 5.2. Of the five results from the two programs, which rule gives the best approximation for a given number of intervals? Also find the *exact* answer using the Fundamental Theorem of Calculus. Show all of your work.

n	2	6	10	30
TRAP				
SIMP				

Chapter 10

Improper Integrals and L'Hopital's Rule

Laboratory Exercise 10.1

Visually Exploring the Graphs of $y = \dfrac{f(x)}{g(x)}$ and $y = \dfrac{f'(x)}{g'(x)}$

Name _____   Due Date _____

Suppose that we have two functions $f(x)$ and $g(x)$ which are differentiable on a certain interval on the x-axis. Let $x=a$ be a point at one end of the interval. Then suppose that both $f(x)$ and $g(x)$ approach zero as x approaches a (because a is at one end of the interval, x approaches it from one side only). The challenge is to find :

$$\lim_{x \to a} \; [f(x)/g(x)]$$

Now make a further assumption that when x is very close to a, the derivative $g'(x)$ has a constant sign (either always negative or always positive). Work through Problem 1 below as a discovery exercise. The graphing calculator screens shown are those for the TI-82. You may need to make viewing rectangle dimensions adjustments depending up the graphing calculator model you are using.

1.   Find $\displaystyle\lim_{x \to 1} \frac{x^2 - 1}{x - 1}$

Enter the numerator, denominator, first derivative of the numerator, first derivative of the denominator, the function in its original form, and the quotient of the numerator derivative and the denominator derivative as shown below. (Note that this last equation *does not* mean the quotient rule for derivatives on $f(x)/g(x)$!) Be sure that when you *turn-off* the first four equations and graph *only* the last two equations.

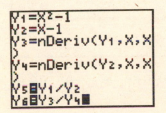 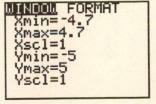 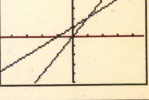

Now TRACE on both Y_5 and Y_6 to the point $x = 1$. What do you see? What do the two graphs *have in common*?

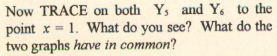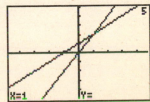 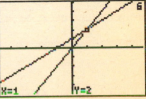

159

2.   Use your graphing calculator to explore the following function and the quotient of the derivative of its numerator and the derivative of its denominator. Sketch your resulting graph on the grid provided below. What do you get for the limit as x approaches zero?

$$\lim_{x \to 0} \frac{e^x - \cos x}{x}$$

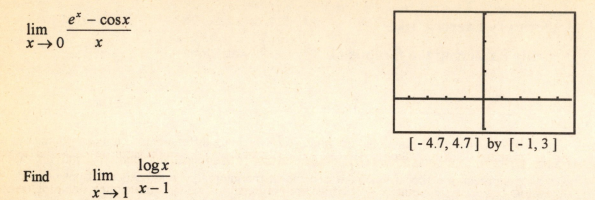

[- 4.7, 4.7] by [- 1, 3]

3.   Find $$\lim_{x \to 1} \frac{\log x}{x - 1}$$

Sketch your resulting graph on the grid provided below. What do you get for the limit as x approaches 1?

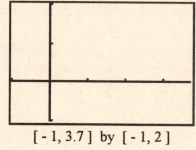

[- 1, 3.7] by [- 1, 2]

By now you should be forming a conjecture. How is that conjecture affected if $x \to +\infty$? Try it out on the following problem.

4. $$\lim_{x \to +\infty} \frac{(\log x)^3}{x}$$

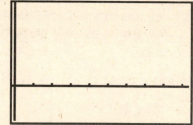

[- 10, 930] by [- 0.1, 0.25]

State in your own words the conjecture, known as L'Hopital's Rule, that you have discovered by graphing the exercise problems.

160

Name _____   Due Date _____

We have used integrals to find the area of a plane region where the lower limit and the upper limit of the integral are finite numbers. What happens when we examine integrals with infinite limits and infinite integrands? Another question we might ask is what happens when a vertical asymptote occurs between the end points of the interval of integration?

As an example, consider the $\displaystyle\int_{-1}^{4}\frac{dx}{(x-1)^2}$. Examining the denominator reveals that a vertical asymptote occurs at the point $x = 1$. If we are not careful when first attempting this problem and do not consider the vertical asymptote, an incorrect answer of -3/2 might be obtained. However, a quick look at the graph of the function itself would prove to be a *warning device* since this vertical asymptote occurs *between* the lower and upper bounds of the integral. This then tells us that *the integral does not exist*!

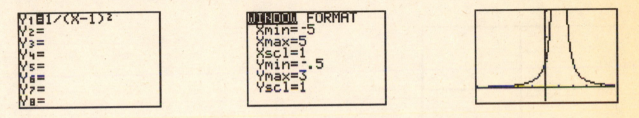

It is possible, however, to evaluate the integral of this function over a *different interval*. That is, we could avoid the vertical asymptote $x = 1$. For example, we could examine $\displaystyle\int_{-1}^{0.99}\frac{dx}{(x-1)^2}$ to get a result of 99.5. Or we could examine $\displaystyle\int_{1.001}^{4}\frac{dx}{(x-1)^2}$ to obtain 999.66666666.

```
fnInt(1/((X-1)^2
),X,-1,.99)
              99.5
fnInt(1/((X-1)^2
),X,1.001,4)
        999.6666666
```

Analytically, evaluate each of the following improper integrals that converge. If an integral fails to exist (see the example above), sketch the graph of the function and then give an example of an interval over which the integral would exist and then evaluate that integral.

1. $\displaystyle\int_{2}^{\infty} \frac{dx}{(x-1)^3}$

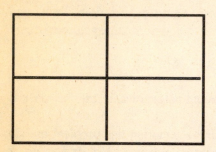

2. $\displaystyle\int_{-1}^{0} \frac{dx}{\sqrt[3]{2x+1}}$

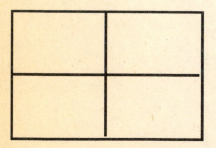

3. $\displaystyle\int_{-1}^{1} \frac{dx}{x^2-1}$

4. $\displaystyle\int_{-\infty}^{0} \frac{dx}{(3x-1)^2}$

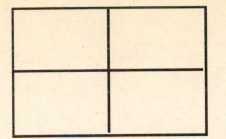

5. Show that $\displaystyle\int_{1}^{\infty} \frac{dx}{x}$ and $\displaystyle\int_{1}^{\infty} \frac{dx}{x+2}$ both diverge.

6. Show that $\displaystyle\frac{2}{x(x+2)} = \frac{1}{x} - \frac{1}{x+2}$.

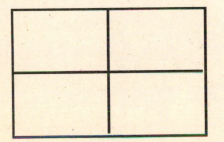

7. Now show that $\int_1^\infty \dfrac{2\,dx}{x(x+2)}$ converges.

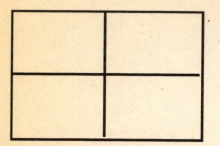

Laboratory Exercise 10.3
L'Hopital's Rule

Name _____   Due Date _____

L'Hopital's Rule tells us that if the $\lim \dfrac{f(x)}{g(x)}$ as x approaches some constant, c, is in one of the indeterminate forms of the kind $0/0$ or ∞/∞ then we can sometimes evaluate the limit by examining the ratio of f' and g'.

Consider the function given below in two different viewing rectangles as *x approaches 2*.

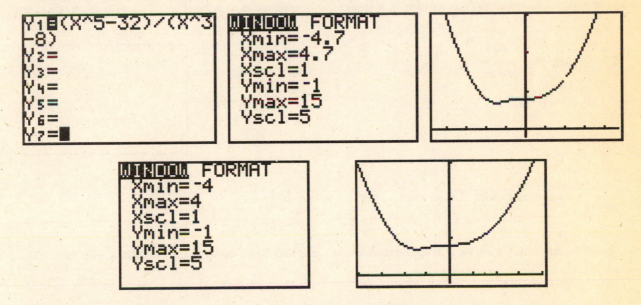

Notice that the first viewing rectangle *does show* that the y-value at $x = 2$ is **undefined**. You can see the gap or empty space in the graph at $x = 2$. However, the second viewing rectangle that differs only in the slightest *does not show or imply this at all!* Be careful in choosing viewing rectangles for the problems in this exercise!! (You may want to refer to Pitfalls of Technology found in the Preface to review how to find friendly screen for your calculator.)

Evaluate the given limit analytically for each of the problems below. Then confirm your analytical results by graphing the original function. Give the dimensions of your chosen viewing rectangle.

1. $\lim\limits_{x \to 0} \dfrac{1 - \cos x}{x^2}$

165

2. $\displaystyle\lim_{x \to 0} \frac{x - \sin x}{2 + 2x + x^2 - 2e^x}$

3. $\displaystyle\lim_{x \to 0} \frac{e^x - \cos x}{x \sin x}$

4. $\displaystyle\lim_{x \to 1} \frac{\pi x}{\cos \pi x}$

5. Analytically find the limit of $\displaystyle\lim_{x \to a} \frac{1}{x - a} \log \frac{x}{a}$.

Support your result with your graphing calculator. Explain how you used the calculator as a support utility.

6. Analytically find the limit of $\displaystyle\lim_{x \to n} \frac{n - x}{\sin \pi x}$, n an integer.

Support your result with your graphing calculator. Explain how you used the calculator as a support utility.

7. Follow the directions for Exercise 10.1 Problem 1 to find: $\displaystyle\lim \frac{e^x - x - \cos 2x}{x^2}$

166

Chapter 11

Infinite Series

Laboratory Exercise 11.1
Sequences and Their Limits

Name _____   Due Date _____

Give the first five terms of each sequence given in the problems below. State whether the sequence converges or diverges. If the sequence converges, find the limit. Support your analytical limit with a graph of the sequence. Also find the **SUM** of the first five terms of the sequence using the **HOME SCREEN** of your graphing calculator and the **2nd LIST** menu. You may need to consult your graphing calculator manual for the correct way to input sequences. The screens below are for the TI-82 graphing calculator. Keep in mind that round-off errors occur when using technology and if calculations are made using rounded-off numbers then the error becomes even larger!

Example: $\left\{\dfrac{n}{n+3}\right\}_{n=1}^{\infty}$

n	1	2	3	4	5
Term Value	1/4	2/5	3/6	4/7	5/8

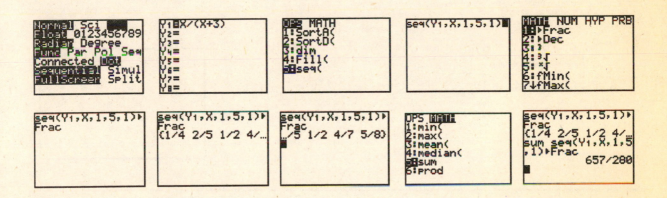

This procedure can also be carried out directly from the HOME SCREEN using a *dummy variable* as shown to the right.

```
sum seq(N/(N+3),
N,1,5,1)▶Frac
            657/280
```

This sequence: **CONVERGES**   **DIVERGES** .

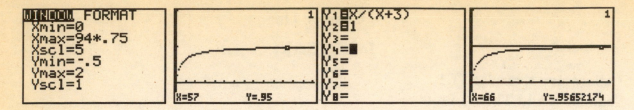

This sequence CONVERGES to a limit of 1.

1. $\left\{\dfrac{n}{n^2+1}\right\}_{n=1}^{\infty}$

n	1	2	3	4	5
Term Value					

This sequence: **CONVERGES TO A LIMIT OF** _____   **DIVERGES** .

SUM of the first five terms = _____

2. $\left\{\dfrac{4n(4n+2)}{(4n+1)(4n+3)}\right\}_{n=1}^{\infty}$

n	1	2	3	4	5
Term Value					

This sequence: **CONVERGES TO A LIMIT OF** _____   **DIVERGES** .

SUM of the first five terms = _____

168

3. $\left\{\dfrac{\ln n}{n+1}\right\}_{n=1}^{\infty}$

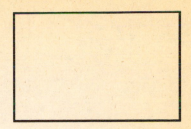

n	1	2	3	4	5
Term Value					

This sequence: CONVERGES TO A LIMIT OF _____   DIVERGES .

SUM of the first five terms = _____

4. $\left\{\dfrac{n}{4^n}\right\}_{n=1}^{\infty}$

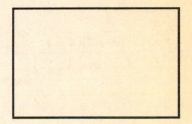

n	1	2	3	4	5
Term Value					

This sequence: CONVERGES TO A LIMIT OF _____   DIVERGES .

SUM of the first five terms = _____

5.　$\left\{\dfrac{n^2-2}{n^2}\right\}_{n=1}^{\infty}$

n	1	2	3	4	5
Term Value					

This sequence:　　CONVERGES TO A LIMIT OF _____　　DIVERGES .

SUM of the first five terms = _____

6.　$\left\{\dfrac{(n+3)(n+5)}{n+1}\right\}_{n=1}^{\infty}$

n	1	2	3	4	5
Term Value					

This sequence:　　CONVERGES TO A LIMIT OF _____　　DIVERGES .

SUM of the first five terms = _____

Laboratory Exercise 11.2
Tests of Convergence

Name _____   Due Date _____

Examine each of the following series and determine which converge and which diverge. Use any of the Convergence Tests that you believe are appropriate for the problem. Indicate which test you used and show all of your work. Give the reasoning for each of your decisions. If the series converges, estimate its sum.

Many graphing calculators have the ability to find the partial sums of a given series. **WARNING!** If the series converges or diverges at a slow pace, you may come to an incorrect conclusion! Some computers/calculators can be very misleading also. A problem such as the following can lead you astray.

$$\sum_{n=1}^{\infty} \frac{1}{\sqrt{n}-4}$$

It is possible that a computer or calculator might give a partial sum for n equal to or greater than 16!! **But** you cannot SUM at or beyond $n = 16$.

1. $$\sum_{n=1}^{\infty} \frac{1}{(n+1)(n+5)}$$

This series: **CONVERGES**   **DIVERGES**   **ESTIMATED SUM:**_____

TEST USED: _____

2. $$\sum_{n=1}^{\infty} \frac{2n}{(n+1)(n+2)}$$

This series: CONVERGES   DIVERGES   ESTIMATED SUM:_____

TEST USED:_____

4. $$\sum_{n=1}^{\infty} \frac{n}{(n+1)(n+3)(n+5)}$$

This series: CONVERGES   DIVERGES   ESTIMATED SUM:_____

TEST USED:_____

5. $$\sum_{n=1}^{\infty} \frac{n+1}{n\ln n}$$

This series: CONVERGES   DIVERGES   ESTIMATED SUM:_____

TEST USED:_____

Laboratory Exercise 11.3
Alternating Series

Name _____   Due Date _____

Test each of the following Alternating Series for divergence, convergence, and absolute convergence. Can you find an approximation for the value of the sum? Then graph the series to 'see' exactly what each looks like. You might try using the **STAT** and **STAT PLOT** features of your graphing calculator if they are available. The screens shown here are from the TI-82 graphing calculator.

Example:

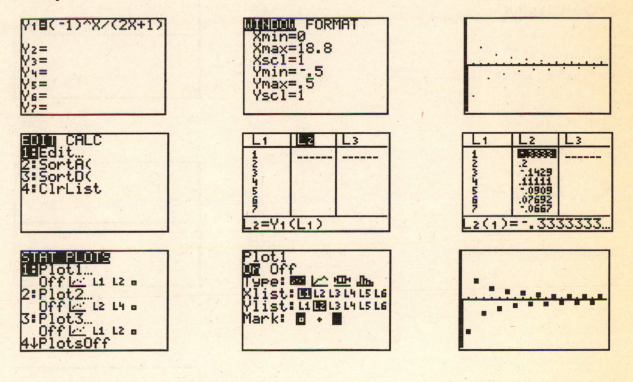

1. $$\sum_{n=1}^{\infty} \frac{(-1)^{n+1}}{n^2+1}$$

2. $\displaystyle\sum_{n=1}^{\infty}\frac{(-1)^n(2n+1)}{n^2}$

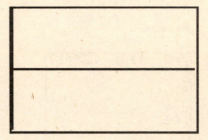

3. $\displaystyle\sum_{n=1}^{\infty}\frac{(-1)^{n+2}}{n\ln n}$

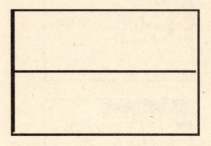

4. $\displaystyle\sum_{n=1}^{\infty}\frac{(-1)^{n-1}}{\sqrt{n}}$

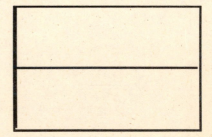

5. $\displaystyle\sum_{n=1}^{\infty}\frac{(-1)^n}{2^n}$

Laboratory Exercise 11.4
Taylor and Maclaurin Series

Name _____   Due Date _____

In each of the first three problems, find the first four Taylor polynomials about $x = a$. Graph the original function along with the four Taylor polynomials on the same grid provided with each problem. Use the **TRACE** feature of your graphing calculator to show that each of the polynomials gives the best approximation of the original function near a.

1.　　$f(x) = 1/x, \quad a = 1$

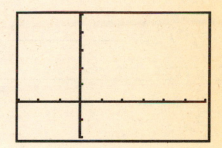

$[-3, 6]$ by $[-2, 5]$

$p_0(x) =$ _____ $p_1(x) =$ _____

$p_2(x) =$ _____ $p_3(x) =$ _____

2.　　$f(x) = \cos x, \quad a = \pi/6$

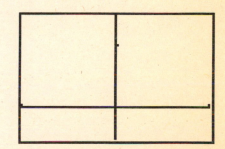

$[-\pi/2, \pi/2]$ by $[-0.5, 1.5]$

$p_0(x) =$ _____ $p_1(x) =$ _____

$p_2(x) =$ _____ $p_3(x) =$ _____

3. $f(x) = x^3 + x^2 - 2x + 1$, $\quad a = 1$

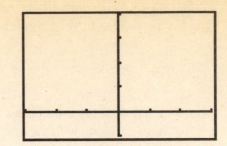

[- 3, 3] by [- 1, 4]

$p_0(x) =$ _____ $p_1(x) =$ _____

$p_2(x) =$ _____ $p_3(x) =$ _____

In each of the next three problems, find the first four Maclaurin polynomials. Show all of your work.
Graph the original function along with the four Maclaurin polynomials on the same grid provided with
each problem. Use the **TRACE** feature of your graphing calculator to show that each of the
polynomials gives the best approximation of the original function.

4. $f(x) = \dfrac{1}{1-x}$

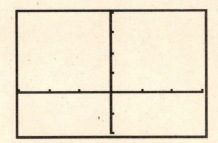

[- 3, 3] by [- 2, 4]

$p_0(x) =$ _____ $p_1(x) =$ _____

$p_2(x) =$ _____ $p_3(x) =$ _____

176

5. $f(x) = \dfrac{1}{1+x^2}$

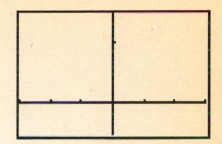

[-3, 3] by [-0.5, 1.5]

$p_0(x) =$ _____

$p_1(x) =$ _____

$p_2(x) =$ _____

$p_3(x) =$ _____

6. $f(x) = \dfrac{1}{(x+1)^2}$

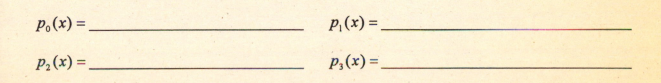

[-3, 2] by [-2, 5]

$p_0(x) =$ _____

$p_1(x) =$ _____

$p_2(x) =$ _____

$p_3(x) =$ _____

Chapter 12

Conic Sections

Laboratory Exercise 12.1
Parabolas

Name _____   Due Date _____

A **parabola** is defined as the set of all points in a plane that are equidistant from a fixed point *(the focus)* and a fixed line *(the directrix)* that does not contain the focus. To the right is an example of the parabola $x^2 = 8y$ with the focus and directrix drawn in. You can easily prove that points on the parabola are equidistant from the focus and directrix by using the Distance Formula. You can also make the parabola open downward by changing the coefficient of y to - 8.

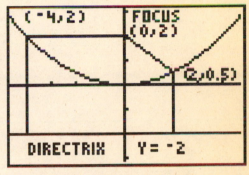

[- 4.7, 4.7] by [- 3, 3]

Graphing a parabola opening to the right or left on the graphing calculator is not as straightforward since the calculator works in **FUNCTION MODE** and the parabola to the right is *not a function*. To 'fake out' the calculator, simply graph $Y_1 = \sqrt{8x}$ and $Y_2 = -Y_1$.

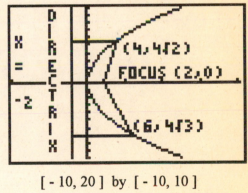

[- 10, 20] by [- 10, 10]

Care should be taken when you use your graphing calculator to see the visual support of the definition of a parabola. If you do not select a ***squared viewing rectangle***, the lengths you are visually comparing could very well 'appear' to be *unequal* instead of *equal*.

Sketch each of the following parabolas on the provided grid. Label the vertex, focus, and directrix. Show all of your work. Then support your graph with your graphing calculator.

1. $x^2 = 5y$

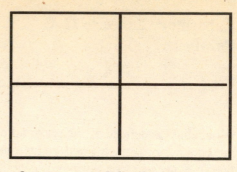

[,   ] by [,   ]

2. $x^2 = -2y$

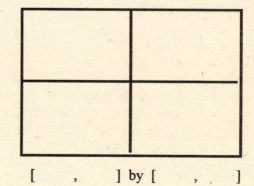

[,   ] by [,   ]

3. $y = x^2 - 8x + 19$

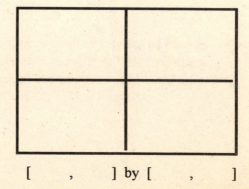

[,   ] by [,   ]

4. $y^2 = -9x$

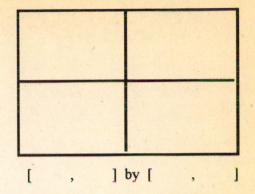

[,   ] by [,   ]

Find an equation for each of the parabolas described below. *Sketch* your result on the provided grid. Support your conclusion by graphing your equation on your graphing calculator. Check the graph to make sure it has the same vertex, directrix, and focus as the problem.

5.   Vertex = (0, 0) and Focus = (0, 5)

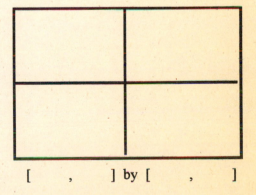

[,   ] by [,   ]

EQUATION: _____

6.   Focus = (- 3, 0) and Directrix is x = 3.

[,   ] by [,   ]

EQUATION: _____

181

7.   Vertex = (- 3, 2)   and   passes through (- 2, - 1).
        Axis parallel to the y-axis.

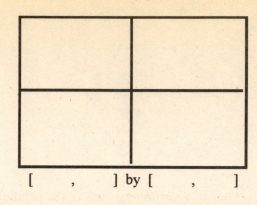

[,   ] by [,   ]

EQUATION: _____

8.   Find the arc length of the parabola $y = 4 - x^2$ from (- 2, 0) to (2, 0).

9.   Find the volume of the solid obtained by rotating the region bounded by $y = 4 - x^2$ and $y = 0$ about the x-axis. Sketch the region on the provided grid below.

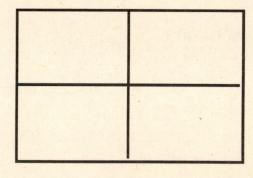

[,   ] by [,   ]

Laboratory Exercise 12.2
Ellipses

Name _____   Due Date _____

Sketch the given ellipse on the provided grid. On the grid label the foci and the end points of both the major axis and minor axis. Show all of your work. Then *support your results* by graphing the ellipse on your graphing calculator. *Remember what you learned in Exercise 12.1 about the viewing rectangle!*

As an example, consider the following ellipse: $\dfrac{x^2}{25} + \dfrac{y^2}{49} = 1$. Solving this equation for y gives

$y = \sqrt{49\left(1 - \dfrac{x^2}{25}\right)}$. In order to graph the ellipse on the TI-82 graphing calculator, *both the positive*

and the negative square roots must be entered into the calculator as is illustrated in the first screen below. The remaining screens show the ellipse in three different viewing rectangles. The window dimensions given here are for the TI-82 graphing calculator. Choose the decimal window appropriate to your calculator model if it is different from the TI-82.

Circle the graph that most __appropriately__ represents this ellipse as we are used to seeing it?

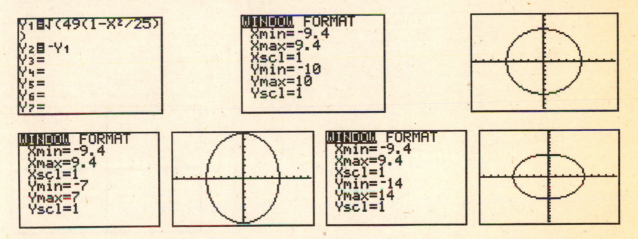

Locate three different points on the ellipse. Use the Distance Formula to find the sum of the distance from each point to the foci. Record your results below.

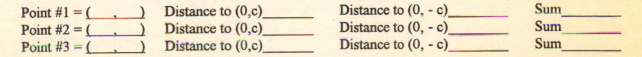

Point #1 = (___ , ___)   Distance to (0,c)_____   Distance to (0, - c)_____   Sum_____
Point #2 = (___ , ___)   Distance to (0,c)_____   Distance to (0, - c)_____   Sum_____
Point #3 = (___ , ___)   Distance to (0,c)_____   Distance to (0, - c)_____   Sum_____

What do you notice about the sums?_____

1. $x^2 + 4y^2 = 9$

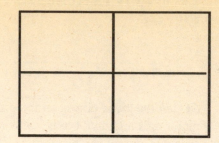

[　,　] by [　,　]

2. $9x^2 + y^2 = 9$

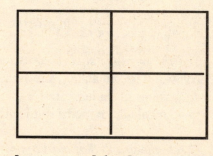

[　,　] by [　,　]

3. $16x^2 + 9y^2 = 144$

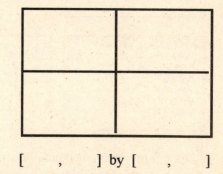

[　,　] by [　,　]

184

4.　　　$4x^2 + 25y^2 = 100$

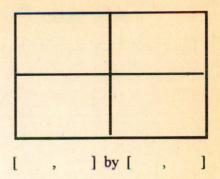

[　　,　　] by [　　,　　]

Find an equation for the ellipse satisfying the conditions given. Show all of your work. *Sketch* the ellipse you find on the provided grid and use your graphing calculator to *support* your results. Label the foci, major axis end points, and minor axis end points. Finally, find *e* of the ellipse.

5.　　　Foci = (± 6, 0)　　　Directrices:　x = ± 8

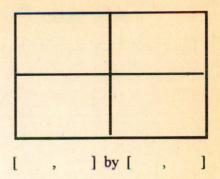

[　　,　　] by [　　,　　]

6.　　　Foci = (0, ± 5)　　　Major axis length :　14

[　　,　　] by [　　,　　]

7. Graph the ellipse: $\dfrac{(x-3)^2}{16}+\dfrac{(y+2)^2}{4}=1$

Label all parts and support with your graphing calculator.

[,   ] by [,   ]

8. Show that $16x^2+25y^2-32x+50y+31=0$ is an ellipse. Graph your results on the provided grid and label all parts. Support with your graphing calculator.

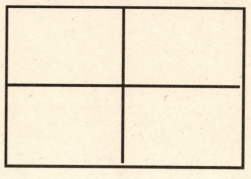

[,   ] by [,   ]

9. Show that $36x^2+9y^2+48x-36y+43=0$ is an ellipse. Graph your results on the provided grid and label all parts. Support with your graphing calculator.

[,   ] by [,   ]

10. Show that $2x^2+2xy+y^2=5$ is an ellipse. Graph your results on the provided grid and label all parts. Support with your graphing calculator.

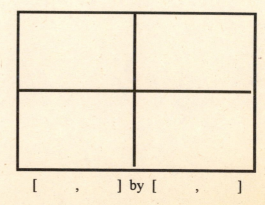

[,   ] by [,   ]

186

Laboratory Exercise 12.3
Hyperbolas

Name _____   Due Date _____

The graph of a hyperbola can be sketched by following these four steps.

    a.   Determine the focus axis.

    b.   Determine the values of a and b and draw a box a units on either side of the center along the focal axis and b units on either side of the center along the conjugate axis.

    c.   Draw the asymptotes along the diagonals of the box you just drew.

    d.   Use the box and asymptotes to sketch in the hyperbola.

As an example the hyperbola $9x^2 - 4y^2 = 36$ is shown below:

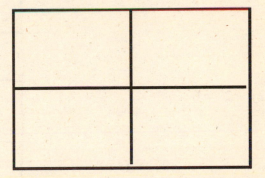

Follow this simple procedure to sketch each of the following hyperbolas. Show all of your work and label all parts of the hyperbola on the provided grid. Confirm your sketching results with your graphing calculator.

1. $4x^2 - y^2 + 16 = 0$

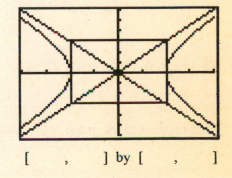

[,] by [,]

2. $36y^2 - 100x^2 = 225$

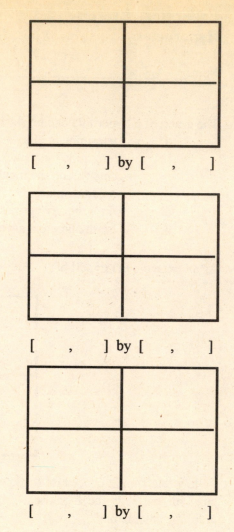

[,   ] by [,   ]

3. $16x^2 - 9y^2 = -36$

[,   ] by [,   ]

4. $3x^2 - 2y^2 - 6x - 12y - 27 = 0$

[,   ] by [,   ]

5.   What graph do you get when you use the Completing the Square method to graph $x^2 - y^2 - 6x + 12y - 27 = 0$?

What happens to the graph if you change the 27 to 26 as follows? $x^2 - y^2 - 6x + 12y - 26 = 0$

What happens to the graph if you change the 27 to 28 as follows? $x^2 - y^2 - 6x + 12y - 28 = 0$

What happens to the graph if you change the 27 to 30 as follows? $x^2 - y^2 - 6x + 12y - 30 = 0$

Laboratory Exercise 12.4
Parametrizations of Plane Curves

Name _____   Due Date _____

Many students find graphing plane curves to be easier (and faster) using Parametric Equations and trigonometric identities.   For instance, let's graph the hyperbola $16x^2 - 25y^2 = 400$ parametrically. To do this we can use the trigonometric identity $\sec^2\theta - \tan^2\theta = 1$ which would result in $x = a\sec\theta$ and $y = b\tan\theta$ where the a and b are those in the standard equation of a hyperbola $\dfrac{x^x}{a^2} - \dfrac{y^2}{b^2} = 1$. Put your graphing calculator into **Parametric Mode**.

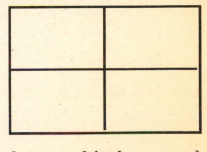

Review your trigonometric identities and graph the following plane curves in **Parametric Mode**. You must first determine if the curve is a circle, parabola, ellipse, or hyperbola in order to know which trigonometric identities to use. Show all of your work, give the parametric equations you use for graphing, and label all parts of each curve.

1. $x^2 + y^2 = 16$ This curve is a _____

[,   ] by [,   ]

Parametric Equations: $X_{1T} =$ _____

$Y_{1T} =$ _____

189

2. $9y^2 + 36x - 6y - 23 = 0$ This curve is a _____

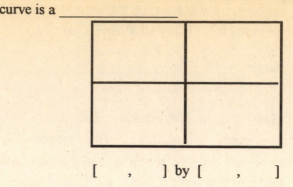

[,   ] by [,   ]

Parametric Equations: $X_{1T} =$ _____

$Y_{1T} =$ _____

3. $4x^2 + 3y^2 - 16x + 18y + 43 = 0$ This curve is a _____

[,   ] by [,   ]

Parametric Equations: $X_{1T} =$ _____

$Y_{1T} =$ _____

4. $9x^2 - 4y^2 - 18x - 24y - 63 = 0$ This curve is a _____

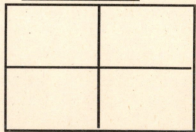

[,   ] by [,   ]

Parametric Equations: $X_{1T} =$ _____

$Y_{1T} =$ _____

Chapter 13

Polar Coordinates and Parametric Equations

Laboratory Exercise 13.1
Exploring Graphs in Polar Coordinates

Name _____   Due Date _____

So far our explorations in graphing have been either in rectangular or parametric form. Now let's explore polar graphs. Exactly what are polar coordinates and polar graphs other than graphing on funny circular graph paper? In this exercise you will explore how to graph familiar curves in polar coordinates instead of rectangular coordinates.

Graph each of the following polar equations and sketch your graph on the provided grid. Be sure that your graphing calculator is in **POLAR MODE**.

1. $r = 5$

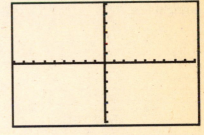

$[-9, 9]$ by $[-6, 6]$
$0 \le \theta \le 2\pi$

What type of graph do you obtain from this polar equation? _____

Would the curve look the same if you used a viewing rectangle of $[-9, 9]$ by $[-9, 9]$? _____

2. $r = 8$

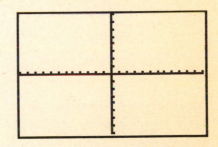

$[-12, 12]$ by $[-8, 8]$
$0 \le \theta \le 2\pi$

State a conjecture about the polar form of the equation of a circle. Then describe the type of viewing window that would produce a circle on your graphing calculator.

191

3. $r = \dfrac{3}{\cos\theta}$

What type of graph do you obtain from this polar equation?

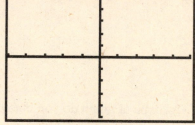

[- 5, 5] by [- 5, 5]
$0 \le \theta \le \pi$

4. $r = \dfrac{-2}{\cos\theta}$

State a conjecture about the polar form of the equation of a vertical straight line.

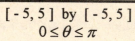

[- 5, 5] by [- 5, 5]
$0 \le \theta \le \pi$

5. $r = \dfrac{4}{\sin\theta}$

What type of graph do you obtain from this polar equation?

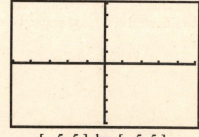

[- 5, 5] by [- 5, 5]
$0 \le \theta \le \pi$

6. $r = \dfrac{-3}{\sin\theta}$

State a conjecture about the polar form of the equation of a vertical straight line.

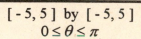

[- 5, 5] by [- 5, 5]
$0 \le \theta \le \pi$

7. $r = \dfrac{1}{\cos\theta + \sin\theta}$

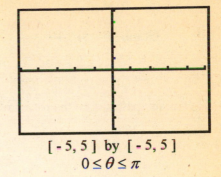

[-5, 5] by [-5, 5]
$0 \le \theta \le \pi$

What type of graph do you obtain from this polar equation?

8. $r = \dfrac{1}{\cos\theta + \sin\theta}$

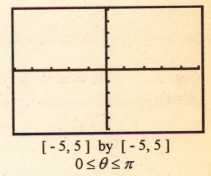

[-5, 5] by [-5, 5]
$0 \le \theta \le \pi$

State a conjecture about the polar form of the equation of a diagonal straight line.

9.   Graph $r = \sin 2\theta$ in the viewing rectangle given below.

You get a four-leaf/petal rose/flower. Why does this happen? When θ is between 90° and 180°, r is negative, thus creating the petal in the fourth quadrant. Between 180° and 270° r is once again positive and the petal in the third quadrant is formed. Lastly, when θ is between 270° and 360° r becomes negative again and the fourth petal is formed in the second quadrant.

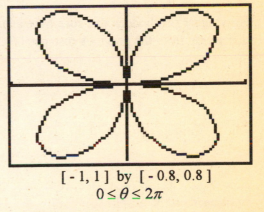

[-1, 1] by [-0.8, 0.8]
$0 \le \theta \le 2\pi$

What happens if the 2θ is changed to 4θ or 3θ or if sine is changed to cosine?

193

10.   Graph $r = \sin 4\theta$.

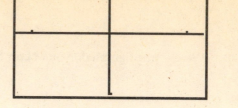

$[-1.2, 1.2]$ by $[-1, 1]$
$0 \le \theta \le 2\pi$

How many petals does the rose now have?

11.   Graph $r = \sin 3\theta$.

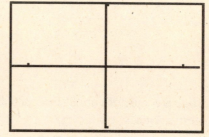

$[-1.2, 1.2]$ by $[-1, 1]$
$0 \le \theta \le 2\pi$

How many petals does the rose now have?

12.   Graph $r = \cos 2\theta$.

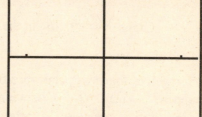

$[-1.2, 1.2]$ by $[-1, 1]$
$0 \le \theta \le 2\pi$

How many petals does this rose have and how are they positioned differently from the previous rose petals?

13.   Graph $r = \cos 2\theta$.

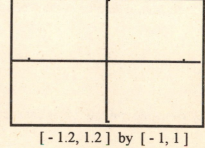

$[-1.2, 1.2]$ by $[-1, 1]$
$0 \le \theta \le 2\pi$

How many petals does the rose now have and how are they positioned on the graph?

Can you state a conjecture about the number of rose petals and their position on the graph from the observations you have made from these graphs?

194

Laboratory Exercise 13.2
Converting Rectangular Graphs into Polar Graphs

Name _____   Due Date _____

Express the following equations in polar form using the identities $y = r \sin \theta$ and $x = r \cos \theta$.
Confirm your results by graphing *both* the original equation in function form *and* its polar equivalent
using the identities for converting rectangular equations into polar equations. Show all of your work
and be sure to give the dimensions of each graph.

1. $y = 4$ Polar Equation:_____

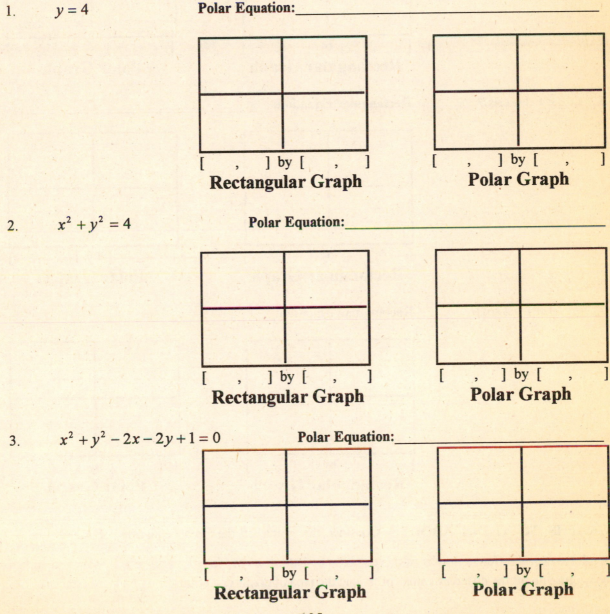

[,   ] by [,   ]   [,   ] by [,   ]
Rectangular Graph   **Polar Graph**

2. $x^2 + y^2 = 4$ Polar Equation:_____

[,   ] by [,   ]   [,   ] by [,   ]
Rectangular Graph   **Polar Graph**

3. $x^2 + y^2 - 2x - 2y + 1 = 0$ Polar Equation:_____

[,   ] by [,   ]   [,   ] by [,   ]
Rectangular Graph   **Polar Graph**

195

Express the following equations in rectangular form. Confirm your results by graphing **both** the original equation in polar form **and** its rectangular equivalent. Show all of your work and be sure to give the dimensions of each graph.

4. $r = \sec\theta \tan\theta$ **Rectangular Equation:**_____

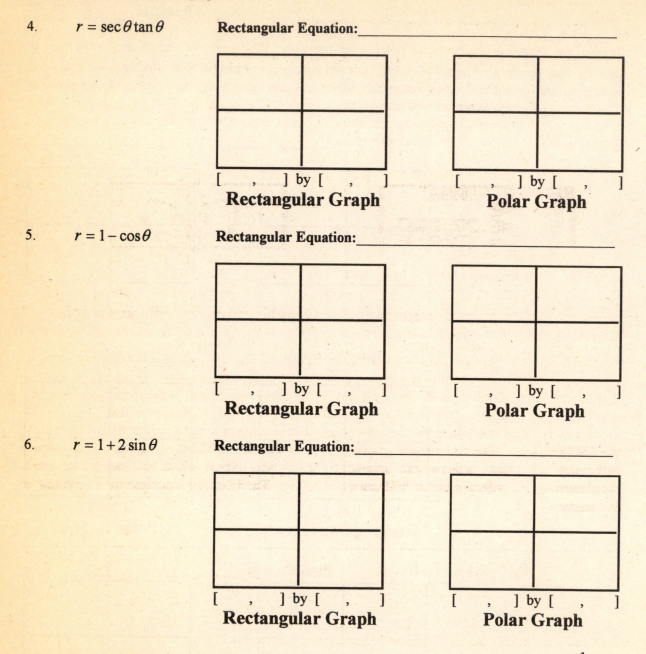

[,   ] by [,   ]
Rectangular Graph

[,   ] by [,   ]
Polar Graph

5. $r = 1 - \cos\theta$ **Rectangular Equation:**_____

[,   ] by [,   ]
Rectangular Graph

[,   ] by [,   ]
Polar Graph

6. $r = 1 + 2\sin\theta$ **Rectangular Equation:**_____

[,   ] by [,   ]
Rectangular Graph

[,   ] by [,   ]
Polar Graph

GRAPHS TO THINK ABOUT: Consider the graph of the polar equation $r = \dfrac{1}{1 - e\cos\theta}$.

Explore several different graphs of this equation by letting $e = 1$, $e < 1$, and $e > 1$. Can you state a conjecture about the different values of e and the type of graph you obtain?

Laboratory Exercise 13.3
Points of Intersection in Polar Coordinates

Name _____   Due Date _____

As with equations in rectangular form, it is possible to solve a pair of equations in polar form simultaneously. However, something very different happens when θ is brought into the graphing process. For example, consider the polar equations: $r = \cos 2\theta$ and $r = \sin 2\theta$. The polar graph of the equations implies that there are eight points of intersection on the petals of the polar rose and one at the origin.

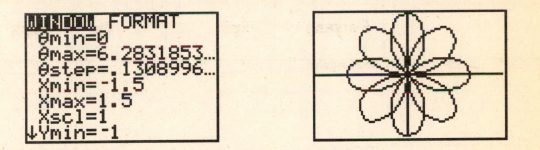

Question: Do all eight points of intersection occur simultaneously? That is, do all eight points of intersection have the *same x, same y, and same θ*? One easy way to answer this question with the help of your graphing calculator is to designate **Simultaneous Mode** when graphing. Then the equations will be graphed *at the same time* instead of the first equation, then the second equation, and so on. Another neat trick that many calculators can do is to *stop* the graphing action by quickly hitting the **ENTER** key. You can resume graphing by pressing the **ENTER** key again. This *turning on/turning off* action process can graphically show you which points of intersection occur simultaneously and which occur at different values of θ. The following snapshots were produced in this manner.

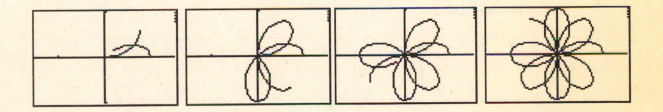

We can now see that four of the eight points of intersection on the rose petals occur at the *same time/θ* ---- once in each of the four quadrants. The other four points of intersection may have the same value

197

for the x and the y terms, but the θ's are different. This can also be verified by using the **TRACE** feature of your graphing calculator.

Find the points of intersection for each of the following pairs of polar equations given below the grids in each problem. Indicate which points, if any, are simultaneous points of intersection. Sketch the graph of each equation separately and then together on the same grid.

1.   **Points of Intersection:**_____

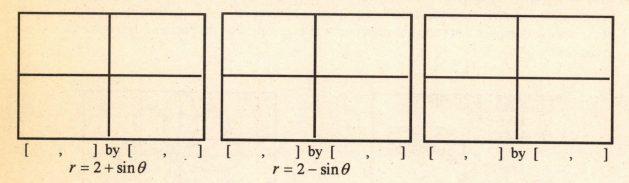

[,   ] by [,   ]   [,   ] by [,   ]   [,   ] by [,   ]
  $r = 2 + \sin\theta$ $r = 2 - \sin\theta$

2.   **Points of Intersection:**_____

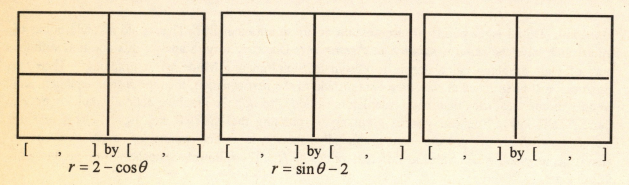

[,   ] by [,   ]   [,   ] by [,   ]   [,   ] by [,   ]
  $r = 2 - \cos\theta$ $r = \sin\theta - 2$

3.   **Points of Intersection:**_____

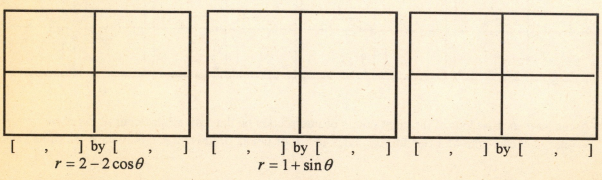

[,   ] by [,   ]   [,   ] by [,   ]   [,   ] by [,   ]
  $r = 2 - 2\cos\theta$ $r = 1 + \sin\theta$

Laboratory Exercise 13.4
Area in Polar Coordinates

Name _____   Due Date _____

1.   Find the area within the inner loop of the graph $r = 1 - 2\sin\theta$. Sketch the graph in order to determine the limits of integration. Shade the region for which you are finding the area. Support your answer with your graphing calculator.

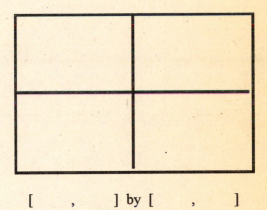

[,   ] by [,   ]

2.   Find the area inside $r = 2 + \cos\theta$. Sketch the graph in order to determine the limits of integration. Shade the region for which you are finding the area. Support your answer with your graphing calculator.

[,   ] by [,   ]

3.   Find the area of the region inside $r = 1 + \sin\theta$ and $r = 1 + \cos\theta$. Sketch the graphs in order to determine the limits of integration and the curves' boundaries. Shade the region for which you are finding the area. Support your answer with your graphing calculator.

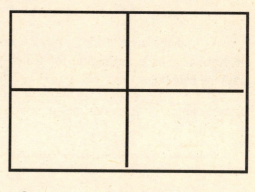

[,   ] by [,   ]

4.   Find the area of the region inside $r = \sin\theta$ and $r = 1 - \sin\theta$. Sketch the graphs in order to determine the limits of integration and the curves' boundaries. Shade the region for which you are finding the area. Support your answer with your graphing calculator.

[,   ] by [,   ]

Laboratory Exercise 13.5
Parametric Equations

Name _____   Due Date _____

1.   Graph the parametric equations $X_{1T} = T^2 - 9$ and $Y_{1T} = \frac{1}{3}T$ for $0 \le T \le 2$ with a T-

step of 0.1 and sketch the result on the provided grid. Indicate on the graph the direction of T as T increases. What happens if the bounds of T are changed? For instance, what if T were changed to $-2 \le T \le 4$? In what way does the graph change? Does the direction of T change if the bounds of T are changed? Sketch this new graph on the second grid and indicate the direction of T as T increases.

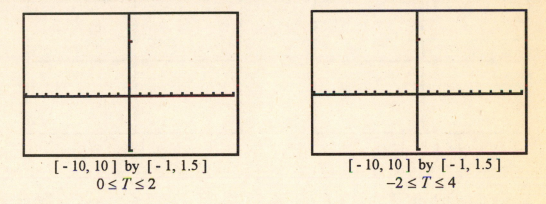

$[-10, 10]$ by $[-1, 1.5]$ $[-10, 10]$ by $[-1, 1.5]$
$0 \le T \le 2$ $-2 \le T \le 4$

2.   Graph the parametric equations $X_{1T} = \sin T + 2$ and $Y_{1T} = \cos T - 2$ for $0 \le T \le 2$
with a T-*step* of 0.1 and sketch the result on the provided grid. Indicate on the graph the direction of T as T increases. What happens if the bounds of T are changed? For instance, what if T were changed to $-2 \le T \le 4$? In what way does the graph change? Does the direction of T change if the bounds of T are changed? Sketch this new graph on the second grid and indicate the direction of T as T increases.

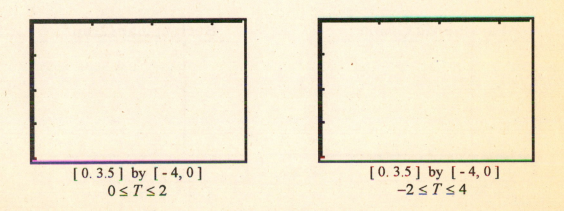

$[0. 3.5]$ by $[-4, 0]$ $[0. 3.5]$ by $[-4, 0]$
$0 \le T \le 2$ $-2 \le T \le 4$

3.   Sketch the graph of $X_{1T} = \sin \pi T$ and $Y_{1T} = \cos 2\pi T$ for $0 \le T \le 2$ with *T-step* of 0.1. Indicate the complete path of the curve as T increases. Then eliminate the parameter T, change the **MODE** of your graphing calculator back to **Function Mode**, and graph the new equation on the second provided grid. Do the graphs match? This is a simple, *fast* method to check your elimination of the parameter T.

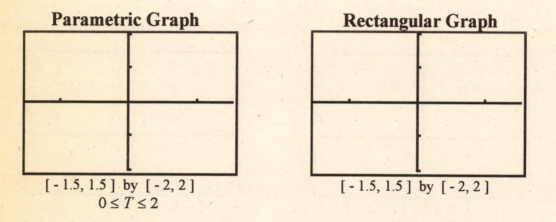

Parametric Graph	**Rectangular Graph**
[- 1.5, 1.5] by [- 2, 2]	[- 1.5, 1.5] by [- 2, 2]
$0 \le T \le 2$	

4.   Sketch the graph of $X_{1T} = T + 1$ and $Y_{1T} = T^2$ for $-2 \le T \le 2$. Indicate the complete path of the curve as T increases. Then eliminate the parameter T, change the **MODE** of your graphing calculator back to **Function Mode**, and graph the new equation on the second provided grid. Do the graphs match? This is a simple, *fast* method to check your elimination of the parameter T.

Parametric Graph

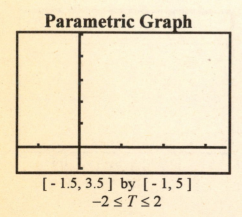

[- 1.5, 3.5] by [- 1, 5]
$-2 \le T \le 2$

Rectangular Graph

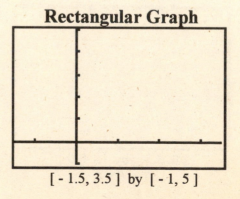

[- 1.5, 3.5] by [- 1, 5]

Laboratory Exercise 13.6
Tangent Lines and Arc Length in Polar Coordinates

Name _____   Due Date _____

1.   Sketch the graph of the polar equation $r_1 = 6\cos\theta$ on the provided graph. When you are
entering the Window dimensions, let θstep $= \dfrac{\pi}{24}$ so that when you **TRACE** the cursor will be in
terms of π for θstep. Then find the slope of the tangent line at the point where $\theta = \pi/4$. Sketch
in this tangent line. If your calculator has a **DRAW TANGENT** utility, you can check the slope that
you obtained analytically with the slope that the **DRAW TANGENT** utility got. What is the equation
of this tangent line? What would be the polar equation for this tangent line? Check your result by
entering the polar equation you obtained into the r_2 position in your graphing calculator and then
graph to see if your tangent line agrees with the tangent line that the calculator obtained. Show all of
your work.

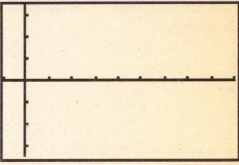

[-1, 9] by [-3.5, 3.5]

2.   Sketch the graph of $r = 2 - 4\sin\theta$ on the provided grid with $0 \le \theta \le \pi$ and θstep $= \dfrac{\pi}{24}$.

On the second grid sketch the graph of r, but let $0 \le \theta \le 2\pi$ with θstep $= \dfrac{\pi}{24}$. What difference, if
any, do you notice in comparing the two graphs? Then find the slope of the tangent line where
$\theta = \pi/4$ and sketch in the tangent line.

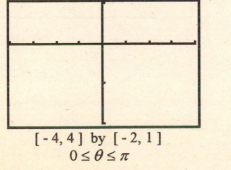

[-4, 4] by [-2, 1]
$0 \le \theta \le \pi$

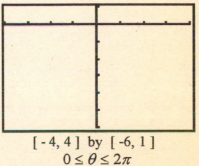

[-4, 4] by [-6, 1]
$0 \le \theta \le 2\pi$

203

3.   Sketch the graph of the polar equation $r = \sin 2\theta$ for $0 \le \theta \le 2\pi$. Find the arc length of the curve from $0 \le \theta \le \pi/2$. Make that section of the curve darker on your graph to indicate the location/position of the arc. Show all of your work.

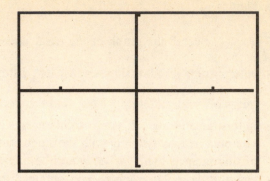

[- 1.5, 1.5] by [- 1, 1]

Can you easily find the arc length of the curve from $0 \le \theta \le \pi$? Explain how you came to your conclusion.

4.   Sketch the graph of the polar equation $r = 1 - 2\sin\theta$ for $0 \le \theta \le 2\pi$. Find the arc length of the curve for $\pi/6 \le \theta \le 5\pi/6$. Make that section of the curve darker on your graph to indicate the location/position of the arc. Show all of your work.

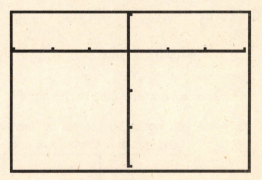

[- 3, 3] by [- 3, 1]